Rethinking Science and Religion in Early Modern Culture

Rethinking Science and Religion in Early Modern Culture

Edited by

Aaron Kitch and Jennifer Rust

THE ARDEN SHAKESPEARE
LONDON • NEW YORK • OXFORD • NEW DELHI • SYDNEY

THE ARDEN SHAKESPEARE
Bloomsbury Publishing Plc, 50 Bedford Square, London, WC1B 3DP, UK
Bloomsbury Publishing Inc, 1359 Broadway, New York, NY 10018, USA
Bloomsbury Publishing Ireland, 29 Earlsfort Terrace, Dublin 2, D02 AY28, Ireland

First published in Great Britain 2026

Cover design: Elena Durey
Cover image © Sergio Azenha / Alamy

A catalogue record for this book is available from the British Library.

Library of Congress Control Number: 2026934014

ISBN: HB: 978-1-350-51162-0
ePDF: 978-1-350-51164-4
eBook: 978-1-350-51163-7

Typeset by Integra Software Services Pvt. Ltd.

For product safety-related questions, contact productsafety@bloomsbury.com.

To find out more about our authors and books visit www.bloomsbury.com
and sign up for our newsletters.

Contents

Figures

Notes on Contributors

Mingna Cheng is a lecturer at Donghua University (CN).

Amy Cooper is Associate Professor of English and Fine Arts at the United States Air Force Academy (CO).

Jacqueline Cowan is an Instructor of English at Red Deer Polytechnic (AB).

Margaret Ferguson is Professor of English Emerita at the University of California, Davis (CA).

Sarah S. Keleher has recently completed her PhD in English from the University of California, Berkeley (CA).

Aaron Kitch is Professor of English at Bowdoin College (ME).

Peter Remien is Professor of English at Lewis-Clark State College (ID).

Jennifer Rust is Associate Professor of English at Saint Louis University (MO).

Jonathan Sawday is Professor of English and the Walter J. Ong, S. J., Chair in the Humanities at Saint Louis University (MO).

A Note on Texts and Spelling

We have modernized spelling and punctuation in all quotations from primary sources. Notes and bibliography follow the 17th ed. of the *Chicago Manual of Style.*

Introduction

Aaron Kitch and Jennifer Rust

Rethinking Science and Religion in Early Modern Culture approaches "science" and "religion" as overlapping and mutually defining modes of thought, belief, and practice in early modern Europe. In addition to reconceptualizing the boundaries between science and religion as practiced and represented in early modern culture, the essays below develop new perspectives on gender and embodiment in considering works and ideas of the past. Some contributors demonstrate how early modern natural philosophers understood their objects of inquiry, from human bodies to the cosmos, in theological terms. Others analyze how religious ideas shaped colonial exploration and the nascent slave trade as scientific enterprises. Still others put early modern poetry and drama in conversation with religious and scientific texts and debates. Across these various inquiries, this volume challenges received narratives of secular disenchantment and liberal progress while questioning conventional assumptions about divisions between scientific and religious communities.[1]

Rather than reduce forms of scientific inquiry to religious pursuits, the chapters below explore shifting and context-dependent boundaries between religious and scientific ideas, beliefs, practices, and modes of representation. Taking up Bruno Latour's invitation to identify new types of scientific and religious "utterance"—while also heeding his warning that scientists and theologians "don't define the true and the false the same way"—the collection develops both new questions and new archives to shed new light on early modern cultures of science and religion.[2] Of course, the categories of "religion" and "science" were (and remain) vast and unsettled, encompassing both individual and collective beliefs, practices, and ideas. Both terms also have the potential to be anachronistic, especially when they appear as abstract nouns. What we would call "scientists" most early moderns would call "natural philosophers," for example, yet the two terms are not equivalent. Natural philosophy typically refers to categories of knowledge derived from Aristotle and often excludes medicine, for instance. And, as Steven Shapin and other contemporary historians of science have argued, early modern natural philosophy is part of a Christian tradition that reads the Book of Nature as a source of spiritual revelation.[3] Before the establishment of scientific disciplines as more formal fields of knowledge in the eighteenth century, the work of science was deeply embedded in older and often overlapping epistemologies, assumptions, and social practices, including religious modes of thought.

At the same time, however, there is some evidence that both "science" and "religion" took on new meanings in early modern Europe. For example, as the *Oxford English Dictionary* records, a new sense of "religion" in its Christian context as a set of abstract beliefs and principles overtook an older sense of "religion" as a group of people living together as part of a particular order of God in the sixteenth century. This is especially evident in post-Reformation examples of "religion" as a synonym for Protestantism.[4] Although multiple definitions of both terms overlapped, "science" also developed a new definition starting around 1600 as a "branch of study that deals with a connected body of demonstrated truths or with observed facts systematically classified and more or less comprehended by general laws, and incorporating trustworthy methods [...] for the discovery of new truth in its own domain." This definition differed from older understandings of science as general knowledge acquired by study.[5] Such a concept of science as a quest for new truths stemming from the empirical study of nature is at the heart of most essays in this collection. Also at stake in many of the works below are the ways in which this new understanding of science shaped religious ideas and practices and was shaped by them in turn, suggesting that "science" and "religion" were actively scrutinized and redefined as categories during the period.

Historical Frameworks: The Conflict Thesis and the Merton Thesis

The "conflict thesis" that posits an intractable and inevitable divide between science and religion is actually the brainchild of nineteenth-century historians. Influential accounts such as John William Draper's *History of the Conflict between Religion and Science* (1874) and Andrew Dickson White's *History of the Warfare of Science with Theology in Christendom* (1896) helped to formulate this misleading dyad, which was also shaped by a tradition of anti-Catholicism stretching back to the Reformation. And yet, however much scholars have attempted to put the conflict thesis to rest, it continues to haunt the contemporary imagination.[6] In his widely read book entitled

God Is Not Great (2007), for example, Christopher Hitchens observes that, "thanks to the telescope and the microscope, [religion] no longer offers an explanation of anything important."[7] This oversimplified claim perpetuates a category error that Terry Eagleton identifies in his response to Hitchens and like-minded atheists: "God for Christian theology is not a mega-manufacturer [...]. Creation is not about getting things off the ground. Rather, God is the reason why there is something rather than nothing."[8] Eagleton's response recasts Hitchens's techno-materialism as an ontological question in a way that is appropriate for thinking about the relationship between science and religion before the Enlightenment. Jacqueline Cowan's chapter below explores how similar contemporary tensions between religion and science uncannily echo early modern debates, as she demonstrates rhetorical affinities between the influential atheist-scientist Richard Dawkins and Thomas Sprat, a prominent framer of the seventeenth-century Royal Society. In their polemical struggles to establish the authority of science vis-à-vis religion, Cowan argues, both Dawkins and Sprat appropriate poetics to the cause of science and attempt to triangulate among seemingly divergent domains.

In contrast to a binary model of conflict, our collection articulates more fluid and mutually reinforcing relationships between religion and science in the early modern period. As John Headley Brooke notes, the conflict thesis portrays science and religion as "hypostatized forces," independent of their concrete and various expressions in specific historical contexts. Echoing Brooke, essays in this volume approach science and religion as "complex social activities involving different expressions of human concern, the same individuals often participating in both."[9] Such an approach favors dialogue over conflict as a model for understanding historically determined relations between science and religion. Though not without tensions, science and religion emerge more clearly as categories that often depend on rhetorical, visual, and embodied media.[10] This volume also rejects a view of science and religion as mirror images of one another, emphasizing both similarities and differences between them, even as they changed in new ways in early modern culture. Authors below address particular relations between science and religion in a range of contexts, acknowledging how claims originating in one scientific or religious sphere can migrate into another and take on new functions and identities. What Thomas Kuhn has defined as scientific "paradigms" change as they take on new content, like old buildings retrofitted for new purposes that ultimately redefine the structures themselves over time.[11]

Understanding science and religion as interconnected social activities rather than monolithic oppositional institutions is the most vital legacy of Robert Merton's much-debated thesis from the 1930s. In his influential work, Merton posited a foundational affinity between "Puritan" values and the emergence of scientific inquiry in seventeenth-century England. Merton essentially extended Max Weber's famous hypothesis about the relationship between the "Protestant ethic" and capitalism to the history of science.[12] In support of his thesis, Merton identified "Puritan" figures among the founding members of the Royal Society alongside other archival evidence. Many elements of Merton's argument have been challenged or qualified by subsequent scholars who have questioned the extent to which some figures in early modern science could actually be considered "Puritans" and the extent to which a "Puritan" social structure was a necessary condition for scientific discovery, given the fact that some of the most groundbreaking discoveries in astronomy in the era by Galileo Galilei, Nicolaus Copernicus, and others arose in Roman Catholic milieux.[13]

The continuing value of Merton's work nonetheless lies in the way that his research offered an alternative approach to the conflict thesis of Draper and others, a point that Merton himself makes in the preface added to his monograph in 1970, where he notes the dominance of the conflict thesis at the time of his original research, a "positivistic" vantage from which his claims about deep relations between religion and science seemed "improbable" and even "absurd."[14] According to Shapin, in the broadest sense, Merton's position has prevailed, whatever the shortcomings of his specific arguments: "no historian of science now seriously contends that religious forces were wholly, or even mainly, antagonistic to natural science. When Merton wrote his thesis, that was not the case."[15] More nuanced versions of the Merton thesis continue to be advanced. Peter Harrison, for example, argues for a more "indirect influence of Protestantism" on scientific development, claiming that Protestant habits of "literalist" biblical exegesis fostered a "mentality" that was conducive to the habits of thought of early modern science.[16] Jonathan Sawday's chapter below, "Counting Time," offers an interesting twist on the question of biblical literalism and science. Sawday demonstrates affinities between the probabilistic methods of seventeenth-century theologians such as James Ussher, who famously dated the age of the earth at around 6,000 years based on scriptural analysis, and those of the Royal Society. Sawday's chapter illuminates the historical contingency and religious significance of large numbers and chronology in the early modern imagination. This focus on seventeenth-century concepts of number contributes to the argument that religiously motivated practices were integral to the rise of scientific inquiry insofar as numbers were essential for creating new forms of knowledge and also frequently charged with theological meaning. Practices such as biblical chronology ironically contributed to the very conditions that allowed for "religion" and "science" to be constituted as opposing fields in modernity.[17] As a whole, the essays in this collection enhance our understanding of how these formations arose in the early modern period.

This collection also bridges two recent "turns" in literary studies that have mostly occurred independently: the "turn" to religion and the "turn" to science. Though most scholars reject the binary opposition between the "two cultures" of science and literature articulated by C. P. Snow in the 1950s, many recent studies of early modern science and literature continue to overlook relevant religious contexts.[18] By contrast, recent work in the history of science has articulated some of the many ways that early modern natural philosophy was enmeshed in religious questions. For example, as Sawday's chapter notes, biblical texts and commentaries were often used as reliable sources of knowledge about the past, present, and future of the divinely created world—the *ordo creationis.* Likewise, theologians adopted the concepts, logic, and methodologies of Aristotelian natural philosophy, elevating scientific activities even as they tried to police the boundaries of orthodox religion. In addition, many religious authorities actually encouraged scientific activity, the persecution of Galileo notwithstanding.

New modes of empirical inquiry into the order of nature beginning in the sixteenth century renegotiated, without necessarily overthrowing, theological foundations of natural philosophy, which for Aristotle and his adherents supported principles of divine design. For example, the influential 1543 study *De revolutionibus orbium coelestium* (On the Revolutions of the Heavenly Spheres) by Copernicus originated as an effort to predict the liturgical calendar more accurately, a topic of great interest to Copernicus as a ranking official of the Catholic Church of Krakow. The church he served did not officially object to his proposed cosmology until 1615, when *De revolutionibus* was added to the *Index Librorum Prohibitorum* at the request of the Dominican priest Giovanni Caccini following the Galileo affair.[19] Assumptions about the role of God as architect of the created world also authorized the experimental and theoretical projects of early modern science. For example, Robert Boyle asserted God's omnipotence as the creator of "nature," while fellow seventeenth-century natural philosopher René Descartes aimed to discover the laws that God has "made with nature" (*composées à la nature*).[20] For Isaac Newton, moreover, gravity was evidence of a divine mover in the cosmos, while the concept of universal space he articulated reflected the omnipotence of a divine creator.[21] Thus, the mechanistic and materialistic approaches to the natural world that we associate today with modern science emerged from within theological frameworks in which natural laws necessarily affirmed the creative power of God.

Adam and the Problem of the Fall in Early Modern Natural Philosophy

The biblical figure of Adam offers another prominent instance of the cross-pollination between religion and early modern natural philosophy that has attracted scholarly interest in recent years. Scholars such as Harrison and Joanna Picciotto have demonstrated how the figure of Adam and the Genesis narrative of the Fall provided a rhetorical *topos* for seventeenth-century natural philosophers to justify new experimental methods and technologies, including the microscope. In these studies, Francis Bacon emerges as a key figure in the seventeenth-century reimagining of the Adamic narrative as a warrant for developing natural philosophy. Bacon cast natural philosophy as a remedy for the impairment of human reason and knowledge after the Fall; his project sought to overcome the "venom which the serpent infused into" human understanding.[22] Bacon and his followers such as Robert Hooke shared the pessimistic view of the Protestant reformers that attributed total depravity to humanity as a consequence of the Fall. In light of the detriments of the postlapsarian condition, knowledge can only be laboriously achieved through testing and experimentation.[23] Thus, the central figure of Adam exemplifies how a theological worldview in the seventeenth century shaped both the goals of natural philosophy (in this case, to repair the defects of fallen knowledge) and its practices of experimentation and technological investigation of the created world.

For Picciotto, the prelapsarian Adam models a productive labor of knowledge not only for Bacon and his "experimentalist" heirs but also for more radical political movements such as Gerrard Winstanley's Diggers. Picciotto's perspective highlights how the development of natural philosophy also had political effects, in this case advancing egalitarian claims and creating a new "public sphere" engaged with the technology and "messy habits of experimentalist observation."[24] Picciotto argues that John Milton exemplifies this shift in the Adamic paradigm in *Paradise Lost* insofar as the poem seeks to redeem "fallen perception," becoming a "medium for a collective revisionary effort" in the spirit of contemporary "experimentalist practitioners" such as Galileo.[25] More recently, Debapriya Sarkar has argued that there are limits to Milton's positive engagement with experimental science, a claim that she pursues interestingly through a focus on Eve rather than Adam, specifically her Fall as a "sad experiment." Sarkar claims that *Paradise Lost* positions the prophetic, disruptive singularity of events like Eve's Fall against the "repeatable experiences" of experimentalism; prophetic events are ultimately "irreducible" to experimental methods in Milton's epic.[26] As these scholars demonstrate, scriptural narratives of the Fall of Adam and Eve were intrinsic to the rise of scientific discourses in the early modern period, whether as a means of engagement or critique.

Several essays in this volume draw out the importance of Adam and Eve to the history of science. While Harrison, Picciotto, and others attribute renewed attention to Adam as a model for knowledge work to the Protestant culture of seventeenth-century England, Sarah Keleher's chapter demonstrates how the figure of Adam in salvation history is crucial to the Catholic Andreas Vesalius's anatomies of the human body in the mid-sixteenth century. Keleher argues that Vesalius's appeals to Adam generate both "anatomical knowledge" and "theological knowledge" through his investigations of the body. Amy Cooper recovers the figure of the "hermetic Adam" in the work of colonialist natural philosopher Hans Sloane, an echo of Bacon's innovative recasting of Adam as a foundational figure for experimentalism. And Sawday's chapter details the importance of the biblical account of Adam and Eve for emergent concepts of chronology in the period.

The problem of the Fall in early modern natural philosophy is not only visible in the paradigmatic figure of Adam, but also in the persistence of negative theology. The medieval *via negativa* represents yet another method for addressing the deleterious effects of the Fall. As Kevin Killeen has recently argued, although mysticism seems to fade after the Reformation, versions of negative theology or "apophatic" rhetoric ("knowing by negation") persist in the inquiries of natural philosophy.[27] Killeen argues that the "apophatic was borrowed in a scientific register of thought" by scholars who grappled with the epistemological detriments of a fallen world.[28] In this case, theology creates new forms of "unknowability" alongside new forms of knowledge, as evidenced in the work of figures such as Boyle and Margaret Cavendish. New terms of knowledge are accompanied by new forms of uncertainty about the boundaries of knowledge of the natural world. Again, several essays in this collection contribute to this line of inquiry. Notably, Peter Remien's chapter builds on Killeen's claims to demonstrate how the methods of negative theology contribute to reflections on natural philosophy by Sir Kenelm Digby and Cavendish. Aaron Kitch's chapter also emphasizes how the idea of the soul presented problems of unknowability for natural philosophers, while Margaret Ferguson draws attention to the gendered dimensions of the unknowable as she focuses on how this problem affects the discourse of the hymen, the elusive organ of the female body that equally preoccupied theological controversies about the Virgin Mary and anatomical investigations.

As these cases illustrate, both "religion" and "science" encompassed multiple and overlapping epistemologies in early modern Europe. For example, Edward Topsell's *History of Four-Footed Beasts* (1607), a translation and adaptation of the Swiss naturalist Conrad Gesner's celebrated study of animals produced in Zürich decades earlier, illustrates a typical mixture of intellectual and religious registers. Like Gesner, Topsell rejected the supernatural accounts of Pliny based on eyewitness investigation, but he also included entries for supernatural creatures such as unicorns and frequently cited the Bible as a source of scientific authority.[29] Gesner and Topsell helped to establish new forms of zoology that drew on empirical observation of animals to challenge conventional biblical truths while also employing religious claims and assumptions in their work.

Theology, Science, and Poetics: Donne's Astronomical Imagination

Literary authors were central figures in the relation between early modern science and religion. John Donne offers a vivid example of not only the epistemological ferment of religion and early modern science in the seventeenth century but also how cultural forms like poetry mediated between these discursive fields. Donne, who was both a poet and a doctor of divinity, showed deep familiarity with early modern scientific developments while also adopting an ambiguous attitude toward them. Mary Thomas Crane claims that Donne expresses dismay at the new philosophy in texts such as *The First Anniversary* (1611), as Donne's speaker famously laments the destabilization of traditional consensus on the nature of the universe.[30] Donne alludes to radical shifts in knowledge about the universe as a source of crisis:

> And new philosophy calls all in doubt,
> The element of fire is quite put out;
> The sun is lost, and th' earth, and no man's wit
> Can well direct him where to look for it.
> And freely men confess that this world's spent,
> When in the planets, and the firmament
> They seek so many new; they see that this
> Is crumbled out again to his atomies.
> 'Tis all in pieces, all coherence gone.[31]

Crane notes that Donne "violates" poetic "decorum" to refer directly to scientific ideas.[32] Like Crane, Howard Marchitello explores Donne's engagement with the prospect of destabilization and his ambivalent stance toward emergent forms of natural philosophy.[33] Donne has long been understood as an interlocutor with emerging forms of science and was likened to "Copernicus" himself by a contemporary.[34] And, indeed, Marchitello argues that Donne participates in the same epistemological revolution that produces the new natural philosophy. In particular, Marchitello claims that, in *Devotions Upon Emergent Occasions* (1624), Donne develops a "*method*" for transforming evidence into universal meaning that shares a "logic" with science, but at the same time seeks to pattern itself on "God's method" as discovered through interpreting embodied experience.[35] Donne emblematizes the complex ways that new forms of empirical investigation can intertwine with theological inquiry in early modern culture.

For example, in his *Devotions*, Donne initially frames Meditation 10 in astronomical terms even as he subsequently subverts a conventional view of microcosm and macrocosm. Donne writes:

> This is nature's nest of boxes: the heavens contain the earth; the earth, cities; cities, men. And all these are concentric; the common center to them all is decay, ruin; only that is eccentric which was never made; only that place, or garment rather, which we can imagine but not demonstrate. That light, which is the very emanation of the light of God, in which the saints shall dwell, with which the saints shall be appareled, only that bends not to this center, to ruin; that which was not made of nothing is not threatened with this annihilation. All other things are; even angels, even our souls; they move upon the same poles, they bend to the same center; and if they were not made immortal by preservation, their nature could not keep them from sinking to this center, annihilation. In all these (the frame of the heavens, the states upon earth, and men in them, comprehend all), those are the greatest mischiefs which are least discerned; the most insensible in their ways come to be the most sensible in their ends.[36]

The referent of the first "this" is uncertain: Donne's disease? His condition? His body? Along with this uncertainty, Donne disrupts the microcosm and macrocosm paradigm. The hierarchies of being, from the heavens to the creaturely world of government and individuals, do not remain ontologically distinct at different levels but rather enfold each other in a "nest." They collapse in on each other rather than maintaining separate but mirroring existences. Their center is "ruin": they all (even the angels) share the condition of creatures created *ex nihilo*, vulnerable to “annihilation” without the “preservation” of divine grace. This tendency toward disintegration spans the cosmos, from the “frame” of heaven to the little world of individual humans. Donne’s argument here resembles his evocation of a world destabilized by the new science in the wake of Galileo and Copernicus in the *First Anniversary*. In such passages, as Marchitello observes, Donne demonstrates an intuition of the “entropic principle *avant la lettre*.”[37] Even though Donne portrays the natural world as inherently fallen, his theological vision emerges in dialogue with an emergent field of natural philosophy.

Donne uses astronomical terms to distinguish between the divine and created worlds. The world collapsing into “ruin” is “concentric”: a term that engenders an image of the orderly celestial spheres of the old Ptolemaic system. Except in this instance, the concentric spheres ultimately orbit nothingness—rather than either the earth or the sun—leading to their eventual decay and disintegration. While Marchitello finds a sneaking “sympathy” for the Ptolemaic version of the cosmos in this passage, Donne might just as well be highlighting the emptiness of this traditional model.[38] It is notable that, in Donne’s scenario, the orbit that disrupts (or at least fails to affirm) the older system is the one that truly manifests divinity. The true “light” of God is “eccentric” to the “ruin” of creation: irregular, decentered from the creaturely sphere, oblique. In Ptolemaic astronomy, “eccentric” referred to a planetary orbit that did not center perfectly around the Earth. By identifying the “eccentric” orb with God, Donne implies that disruptions in classical astronomical knowledge that might be associated with the new philosophy are not necessarily antithetical to divine revelation. However, “eccentric” also had a use in the Copernican system, as it could equally refer to an orbit that defied heliocentrism.[39] In associating divinity with astronomical “eccentricity,” Donne suggests that it may always stand at an oblique angle from any human system of knowledge, even as it renders the created realm possible. Astronomy, a scientific field whose upheaval was well known to Donne, offers a vocabulary for the pessimistic vision of a fallen cosmos posited at the outset of this Meditation, again demonstrating how Donne’s literary imagination merges theology and new science, as well as the ways that this imagination can deviate from modern expectations about scientific progress.

As this passage from Donne’s *Devotions* illustrates, poetic forms engaged with both religious and scientific discourses, not simply reflecting these discourses but offering imaginative alternative epistemologies of their own. Essays by Cowan and Mingna Cheng in the third part of the collection focus in particular on some of the ways that forms of “poesy” interacted with religion and science in the seventeenth century and later periods. The case of Donne also reveals the ways that important intellectual frameworks of earlier centuries continued to exert pressure even in a period of significant scientific innovation. As Edward Grant notes in his sweeping study of *Science and Religion* from 400 BCE to 1550 CE, for example, theologians between 1200 and 1500 elevated the prestige and reach of natural philosophy by adapting Aristotelian methods and logic to theological questions while also working to limit the theological range of natural philosophy. The more rationalized and logical forms of theology articulated by Aquinas and others in the twelfth century who benefitted from newly translated texts of Aristotle became a foundation for curricula in universities around Europe beginning in the thirteenth century. These institutions, which were themselves of relatively recent origin, in turn helped to elevate theology as the most prestigious field of study, yet this new prestige also created new pressures to police the boundaries of belief. For example, the "Condemnation of 1277" authored by the Bishop of Paris, Stephen Tempier, sought to restrict propositions and lines of questioning related to the knowability of God. Yet, as Grant notes, this historical shift had the unintended consequence of allowing natural philosophers to pursue new questions about the divinely created natural world so long as they did not contradict basic principles of theology. Influential natural philosophers such as William of Ockham and John Buridan responded to this opportunity by elevating Aristotelian natural philosophy while carefully avoiding controversial propositions about the nature of God.[40] This dynamic shifted again during the Reformation as relations between the state and the church also realigned, allowing more expansive scientific inquiry even amid the backlash of the Inquisition.

New visual technologies such as the microscope and telescope together with the widespread adoption of the printing press also facilitated innovative experiments and accounts that many still celebrate as a "Scientific Revolution." Prominent examples include Vesalius's study of human anatomy, Galileo's observation of moons orbiting Jupiter, William Gilbert's theory of magnetic fields, and William Harvey's discovery of the circulation of blood in the human body. As different as these and other discoveries were in their scientific methodology and truth claims, all sought to study "nature" more directly. Harvey proudly declares in the introduction to his *De Motu Cordis* (Anatomical Exercises Concerning the Motion of the Heart and Blood in Living Creatures, 1628) that his knowledge of the human body derives "not from books but from dissections, not from the tenets of Philosophers but from the fabric of Nature."[41] The scientific discoveries enabled by this concerted return to nature subsequently put pressure on older epistemologies, which in turn put old tropes to new uses. As Harrison argues, for example, the Book of Nature shifted in meaning during the sixteenth century in response to new forms of empiricism, even as foundational assumptions about the divine origins of nature (the *ordo creationis*) remained in place.[42] Another trope also emerged at this time: the "theater" of nature, which was developed most fully in early modern Europe by the French legal theorist Jean Bodin. Bodin's work foregrounds God's providential government within the study of nature and outlines a more dynamic natural order marked by sophisticated forms of representation and a metaphorics of public performance.[43] This trope together with others in the period speaks to the way that the "back to nature" movement of the sixteenth century was sometimes paradoxically dependent on forms of representation in order to produce new knowledge.

Francis Bacon and the Book of Nature

Bacon's ambitious efforts to reform and expand knowledge practices and fields of natural philosophy likewise illuminate the foundational intersection of science and religion in early modern England, as several contributors to this volume acknowledge. Bacon has understandably been the subject of much inquiry because of his enormous influence on scientific thought and activity in seventeenth-century England, but less attention has been paid to his religious beliefs.[44] Raised by an adamantly Protestant mother (Anne) and educated by a Puritan tutor (John Walsall), Bacon also attended Cambridge at a time when it was a hotbed of Calvinism. Although he sought at times in his writing to differentiate between natural philosophy and theology, he explicitly framed his reformation project as a moral and ethical undertaking. While his embrace of inductive philosophy as the most effective way to study the Book of Nature did not contradict his fundamental faith in nature as a product of divine creation, he did try to limit the kind of theological knowledge that could be derived from the study of nature. Like many of his contemporaries, moreover, Bacon embraced a conventionally Christian mode of natural theology that assumed a divine author of the *ordo creationis*, noting in *The Advancement of Learning* (1605) for instance that "all works" of nature illustrate "the Omnipotency and wisdom of the Maker" and reveal God's "power, providence, and goodness."[45] But in urging natural philosophers to examine the Book of Nature in light of empirical methods, he also urged limits on the kinds of theological conclusions that they drew from the study of nature, in part due to his interpretation of the negative effects of the Fall on human understanding. In this sense, Bacon proposed more limited forms of natural theology than many of his contemporaries, though ironically the more organized forms of scientific activity that he helped to spur in England themselves became sources of new theology after his death. The field of "physico-theology" that emerged in late seventeenth-century England, for example, mixed empirical methodologies, natural history, and Newtonian physics to form a new kind of natural theology grounded in the empirical study of nature.[46] As Rosalind Powell has argued, "physico-theology" profoundly shaped the theories of perception that were woven into articulations of "Newtonian popular science" into the eighteenth century.[47]

Bacon followed numerous Christian natural philosophers of medieval and early modern Europe who drew on the Augustinian trope of the Book of Nature to justify forms of empirical inquiry. But, as Mingna Cheng remarks in her chapter below, Bacon reimagined natural philosophy beyond Aristotelian traditions to develop a new relationship with the order of nature. Even though he was skeptical about the human capacity to discern the presence of God in the natural world, Bacon also viewed natural philosophy as an eschatological project.[48] For example, he alludes directly to Daniel 12:4 ("But thou, O Daniel, shut up the words, and seal the book, even to the time of the end: many shall run to and fro, and knowledge shall be increased" [King James Version]) as prophetic evidence for a time when new knowledge would flourish—as evidence that God himself sanctioned scientific inquiry.[49] Although he invoked the Book of Nature, as Cheng notes, Bacon also established nature as more autonomous and external to human existence, making it the subject of scientific examination and manipulation in new ways.

Bacon's life and works thus provide one example of the complex interweaving of science and religion in early modern Europe—at once complementary and antithetical, foundational and oppositional. Relations between theology and natural philosophy also shifted in response to specific political, religious, and cultural influences, including the category of imaginative writing that Philip Sidney called "poesy." Bacon's approach to poesy differs from Donne's, for example, although they both share a similarly pessimistic view of the effects of the Fall. Bacon's unfinished utopian narrative of *The New Atlantis* (1627) offers one example of how science and poesy intersect, as Cheng explores below. Bacon's fascination with literary form joined his interests in rhetoric, history, and poesy. In its broadest sense as literary and rhetorical form, poesy was also central to Bacon's metaphysics as a medium that encompassed relations between the fallen human imagination and the *ordo creationis*. Bacon defined both the mind and the world it reflects in terms of literacy and forms of representation, seeking famously to make "experience" itself "literate."[50] By learning its "letters," Bacon hoped, natural experience could leave behind the particular and private concerns of individuals in order to reach the public sphere, where it could generate new modes of collective knowledge. Such advocacy helped push "experiment" from its older meaning of "experience" toward the more modern sense of testing a hypothesis using scientific instruments in controlled settings.[51]

Rejecting the foundational distinction that Aristotle had made between art and nature, Bacon saw them as mutually dependent—both could be elevated to the condition of *scientia* or scientific knowledge. As he argued in the second book of *The Advancement of Learning*, it should be the "duty of Art to perfect and exalt Nature."[52] Bacon also famously urged experimental natural philosophers to test or "vex" nature and suggested that mechanical arts, or what he called the history of "nature wrought," should be included in the study of natural history. He noted that "nature cannot appear so fully in the liberty of Nature, as in the trials and vexations of Art."[53] In advocating for new systems of knowledge, Bacon recognized the scientific value of art in many forms not only as instruments of knowledge but also as modes of expressing divine creation in human language. As Cowan argues below, moreover, forms of fiction were essential for this project, and Bacon echoed earlier defenders of poesy such as George Puttenham, who describes poesy as "noble, profitable, ancient, and divine [...] science."[54] Bacon's stance emphasizes continuities between poetic art and emergent scientific technologies, both forms of human creation that ultimately also affirm an immanent divine order.

Many of the essays in this collection trace Bacon's complex scientific, theological, and aesthetic legacies through an array of discourses. Our contributors often demonstrate how early modern poetic and scientific discourses position themselves vis-à-vis the Royal Society, for example, which authorized itself in Baconian terms and claimed Bacon as its progenitor. Several contributors also show how encounters between early modern science and religion go beyond Baconian paradigms, extending, for example, to complex concepts of embodiment and anticipating contemporary reframings of nature as ecology and controversies between advocates and critics of purportedly atheistic "science" against the specter of "religion."

Overview of Chapters

We have divided the chapters in this volume into three parts according to their central claims and concerns.

Part One: Divine Matter: Embodiment in Early Modern Science

The three essays in Part One all recognize how questions of embodiment in works of natural philosophy were thoroughly invested in religious questions. The body was not simply an object of natural study but also a vehicle of devotion and an exemplary work of creation within a cosmos governed by divine principles. Sarah Keleher's opening chapter on Vesalius rejects claims of scientific secularization that often surround his revolutionary study of anatomy by identifying a central narrative of salvation history in his work. Vesalius in his *De humani corporis fabrica libri septem* (On the Fabric of the Human Body in Seven Books, 1543) highlights God's formation of the Adamic body and extends through the bodily corruption of the Fall and Christ's redemptive embodiment in the Incarnation. Yet this soteriological structure generates a seemingly intractable problem, since the fallen anatomist cannot directly access the prelapsarian body, the living and ensouled body, or what Keleher calls "the resurrection body." To solve this problem, Vesalius argues that the soul and body eventually reunite to generate fuller anatomical knowledge in the name of Christ himself. Such a purified resurrection body provides a *telos* of anatomy as a science for Vesalius, defining the aspirational perfection of the body that also provides full anatomical knowledge to the observer. In her conclusion, Keleher addresses the importance of this soteriological body for the nascent science of anatomy in early modern Europe, outlining a new vision of the inextricable bonds between science and religion in the period.

Like Keleher, Margaret Ferguson focuses on the body as a site of both anatomical and theological significance. Her chapter, "Strange Evidence: Hymens Between Science and Religion," raises questions about what counts as visual evidence in both theological and scientific domains. Ferguson argues that the hymenal membrane represents not just a somatic sign of virginity but also a contested site of bodily and religious knowledge in early modern discourses that range from theological disputes about the perpetual virginity of Mary to anatomists' attempts to establish this elusive bodily sign of sexual purity. This construction of the female body challenges Renaissance theories of vision by positing the hymen as the uncertain yet visible sign of the sanctity of virginity. Focusing on visions of virginity in a range of texts, including Shakespeare's *As You Like It* (*c.* 1599), Ferguson articulates the hymen as a challenge to truth-claims about virginal female bodies as reflected in a range of early modern religious and scientific texts. Her chapter acknowledges the distinct epistemologies of early modern science and theology while foregrounding central questions of gender and embodiment that both domains share.

Aaron Kitch addresses different modes of embodiment in relation to the soul, an object of both scientific and religious inquiry in classical, medieval, and early modern Europe. His chapter on *Hamlet* traces the soul as a problem of both knowledge (epistemology) and being (ontology) in works of philosophy, natural philosophy, and theology. Shakespeare's representation of the soul in his famous tragedy blends spiritual concerns with questions of embodied desire, identity, and representation or mimesis, following accounts of Aristotle and others. His reading of *Hamlet* identifies spiritual dimensions that differ from those outlined by Stephen Greenblatt and Peter Lake and that modify Rhodri Lewis's Ciceronian reading of the soul.[55] Ultimately, Kitch shows that Shakespeare registers new definitions of both "soul" and "spirit" as modes of troubled subjectivity, drawing on Donovan Sherman's claim that the soul in *Hamlet* is "a spiritual form that occupies the contact point between knowing and not-knowing" to locate the soul in the play as a profoundly ambiguous source of cognition, embodiment, and selfhood.[56] *Hamlet* depicts the soul in ways that draw on central arguments of premodern and early modern theology and natural philosophy, an intellectual context that also allows the play to raise central questions about the capacity of the stage to represent human identity.

Part Two: Legacies of Early Modern Science and Theology

The essays in Part Two explore theological traditions and biblical allusions that helped early modern natural philosophy to define new conceptions of nature and new horizons for scientific exploration, including modes of inquiry that helped make way for the colonization of the non-European world. Contributors demonstrate how sometimes unrecognized theological undercurrents inform frameworks of scientific understanding that continue to have resonance today insofar as they undergird modern theories of ecology, scientific legacies of colonialism and slavery, and large-scale numbers and conceptions of chronology. In "Versions of Negative Theology: Kenelm Digby and Margaret Cavendish," Peter Remien argues that the natural philosophical systems of Digby and Cavendish offer contrasting visions of how the new science of the seventeenth century could be used in service of negative theology. Digby and Cavendish were among the first English writers to engage with the mechanical philosophy of Continental philosophers such as Descartes and Pierre Gassendi. Digby's *Two Treatises* (1644) was the first comprehensive work of mechanical philosophy published in English. Digby's aim was primarily theological in nature: to use the epistemological tools of the new science to prove the immortality of the soul. The natural philosophy of Digby's friend Cavendish, Duchess of Newcastle, encompassed a different version of negative theology based not on Cartesian dualism but on Nature's adjacency to God. For Cavendish, God and Nature are coextensive, though not equal, and Nature is eternally creative rather than simply created, irreducible to inert atoms or mechanical principles. Cavendish's attempts to understand Nature and the God of Nature are empathetic, imaginative, and collaborative rather than empirical and rational like Digby's vast mechanical system. Remien ultimately argues that the negative theologies of these early modern natural philosophers anticipate intellectual currents in contemporary ecological studies. Where Digby's ecology rests on a predictable and knowable nature that can be quantified empirically, Cavendish envisions a radical separation of human knowledge from ecology that prefigures negative ecology by embracing the complexity of the world.

Amy Cooper considers a different theological legacy of Baconian science in her chapter on the Irish collector, physician, and colonialist Hans Sloane. Admitted to the Royal Society in 1685, Sloane undertook a comprehensive study of medicinal plants in Jamaica, where he also treated hundreds of patients. Cooper situates Sloane within expansive networks of colonial trade and theology, noting that both Bacon and Sloane embraced the biblical King Solomon as a natural historian: Bacon elevates Solomon as a model of scientific inquiry in *The New Atlantis* and praises King James I as a new Solomon in his dedication to *Instauratio Magna* (1620). This framing of Solomon as a scientist, Cooper maintains, mobilizes the Hermetic tradition that was rediscovered for a European audience in the fifteenth century by Marsilio Ficino. In fashioning himself as a Baconian natural historian—and indeed, as the New Solomon that Bacon envisions in the *Instauratio*—Sloane places himself in an intellectual genealogy that is both biblical and Hermetic in origin. Cooper argues that Bacon's Hermeticism is also racialized: the ship sailing through the Pillars of Hercules on the frontispiece of the printed editions of Bacon's *Novum Organum* (1620) asserts the colonial ambitions of Baconian science and may resonate visually with the slave trade. Her reading of Sloane backward through Bacon to the Hermeticism of classical antiquity thus offers a new chapter in the history of British colonialism in which scientific and spiritual concerns commingle.

Jonathan Sawday considers another early modern legacy of religion and science in his chapter on the problem of temporality and large-scale numbers in post-Reformation scientific and theological reasoning. Sawday focuses in particular on the attempts of Joseph Scaliger and James Ussher to reconcile new concepts of measurable and quantifiable time with biblical accounts of creation in a way that was compatible with human history. Ussher's project, Sawday argues, was comparable to the work of renowned Renaissance astronomers such as Johannes Kepler. But Sawday also attends to Ussher's notable afterlife in twentieth-century American creationism, where it offers a counter-logic to modern science that also deals in the language of evidence and chronology. Sawday observes how entities such as the Creation Museum claim to support Ussher's hypotheses about the age of the earth but also generate a deeper and overlooked irony in the fact that Ussher sought to synchronize his efforts to identify "measureable" time with others engaged in early modern science, including members of the Royal Society, while his heirs in the movement of Young Earth Creationism maintain a rigid stance against modern science in a hyperbolic version of the "conflict thesis." Sawday associates these creationists with Milton's Satan in *Paradise Lost*: "Both he and they are convinced of their essential changelessness, as one wages war against God, and the other against modernity." From this perspective, however, the triumph of modern science also seems less certain, as seventeenth-century forms of knowledge persist to haunt contemporary discourses of biology.

Part Three: Poesy and the Mediation of Religion and Science

Essays in Part Three examine how poetics often mediate differences and tensions between theological claims and the practice of natural philosophy. While eschewing the extreme oppositionality of the conflict thesis, these essays examine the interplay between poetry, science, and religion, indicating that "antagonism" can be as "generative a form of engagement as concord," as Elizabeth Swann has recently argued about the relationship between "poesy" and "scientia" in the period.[57] The appropriation of poetics to reinforce the religious authority of early modern science has a legacy that lingers today in polemical efforts to harness poetry to science so as to override the irrational persistence of religion.

Mingna Cheng addresses the poetics of science in connection with the Royal Society, both in terms of the impact of Baconian science on the poets Abraham Cowley and John Dryden and in the development of a gendered poetic metaphor describing the relationship between science and nature. Focusing on poems written by Cowley and Dryden to honor the Royal Society, Cheng analyzes divergent expectations about the relationship between nature, poetry, and science among the three authors. Where Bacon in his *Advancement of Learning* describes natural philosophy as a "spouse" for human "generation, fruit, and comfort," Cowley employs a trope of nature as "heir" and Dryden a language of nature as "handmaid." Cheng demonstrates how the figure of Bacon and the related activities of the Royal Society generated varying poetic responses in the late seventeenth century. Tensions between the theological and materialist aspirations of Baconian science and its fraught relationship with the poetic arts became visible in the engagements of Cowley and Dryden. While Cowley seeks to advance the project of Baconian science through innovations in "empirical" poetics, Dryden adopts a more skeptical stance in *Annus Mirabilis* (1667) in which poetry can complement but also resist the claims of mechanistic natural philosophy.

Jacqueline Cowan's chapter explores a different question about poetry, religion, and the Royal Society by constructing a transhistorical dialogue between its founder Thomas Sprat and the modern-day scientist and prominent atheist Richard Dawkins. Cowan focuses on the ways that both Sprat and Dawkins invoke poetics in order to navigate tensions between religion and science. Although Sprat was a devout member of the Anglican Church and Dawkins is a stalwart atheist, each betrays anxiety about the deleterious effect of inflammatory rhetoric and religious passion. Sprat adopts a poetic imagination to frame experimental philosophy as a return to what Royal Society Fellows considered the original form of worship, where Dawkins hopes to diminish the authority of religion itself. Sprat appropriates the claims of Renaissance poetics to identify the "golden world" of poetry with the world revealed by experimental science, while Dawkins offers poetics as a solution to the supposed incommensurability between literature and hard science. But both take refuge in the capacity of poetry to lessen religious conflict. For Dawkins, poetic science is a superior replacement to what he considers to be the irrational enchantments of religion. Yet the strategy that Dawkins employs to separate science from religion turns out to be the very same that Sprat used to usher experimental philosophy back into the good graces of theologians. An irony emerges in the sense that scientists, both past and present, leverage poetics to bolster the authority of science while also devaluing poetic language, even if such language knits science, religion, and literature closer together.

Together, these essays seek to redefine both early modern religion and science as fluid and expansive discourses in which the rhetoric of the past continues to return uncannily within the controversies of the present. Science and religion were and remain mutually and dialogically constituted as forms of knowledge, belief, practice, and power. The wide-ranging contributions to this volume return to fundamental questions and assumptions that underwrite both fields while also working to establish new questions, archives, and intellectual genealogies.

Notes

1 For a compelling, surprising, and statistically focused study of "shared efforts between" contemporary "scientific and religious communities," see Elaine Howard Ecklund and Christopher P. Scheitle, *Religion vs. Science* (quote on 145).

2 Latour, *Rejoicings*, 27.

3 See, e.g., Shapin, *The Scientific Revolution,* 3–4 et passim.

4 *Oxford English Dictionary* (*OED*), "religion (n.)," def. 4.a and 4c.

5 *OED*, "science (n.)," def. 4.a and 2.

6 For histories of science that engage the shaping influence of religion, see, e.g., Harrison, *The Bible, Protestantism, and the Rise of Natural Science* and *The Fall of Man and the Foundations of Science*. The historian of alchemy Lawrence Principe also argues that "theological motivations" were the "single greatest driving force for scientific inquiry throughout the entire early modern period." See *The Scientific Revolution,* 57.

7 Hitchens, *God Is Not Great*, 282.

8 Eagleton, *Reason, Faith, & Revolution*, 7.

9 Brooke, *Science and Religion*, 56.

10 For more on the model of dialogue, see Ian G. Barbour, *When Science Meets Religion*, 2–3.

11 Kuhn developed his theory of scientific "paradigms" most fully in *The Structure of Scientific Revolutions*. Although Kuhn's theory of scientific change has received criticism, especially in his notion of the "incommensurability" between paradigms, his conclusion that science does not progress in a linear fashion toward greater "truths" continues to shape studies in the history of science. On Kuhn's legacy, see *Kuhn's Structure of Scientific Revolutions—50 Years On*, ed. Bokulich and Devlin.

12 Merton, *Science, Technology and Society*; Weber, *The Protestant Ethic and the Spirit of Capitalism*.

13 For critiques of Merton, see Harrison, *The Bible, Protestantism, and the Rise of Natural Science*, 5–6; and Shapin, "Understanding the Merton Thesis." On the support by the Catholic Church of early modern scientific inquiry, see Ronald L. Numbers, *Galileo Goes to Jail*, esp. 100–3.

14 Merton, *Science, Technology and Society*, xvi.

15 Shapin, "Understanding the Merton Thesis," 604.

16 Harrison, *The Bible, Protestantism, and the Rise of Natural Science*, 8.

17 Harrison identifies a similar dynamic in Reformation-era reinterpretations of Aristotelian causality; see *Territories of Science and Religion*, 80.

18 Examples of works of literature and science that marginalize religion include Elizabeth Spiller's *Science, Reading, and Renaissance Literature*; Mary Thomas Crane's *Losing Touch with Nature*; and the essays gathered in the *Palgrave Handbook of Early Modern Literature and Science*, ed.

Marchitello and Tribble. Likewise, essays on early modern religion and literature such as those collected in *Shakespeare and Religious Change*, ed. Graham and Collington, approach religious topics without considering early modern science/natural philosophy. More recently, some approaches to science in Shakespearean drama have addressed the role of the supernatural, including Kristen Poole's *Supernatural Environments in Shakespeare's England* and the recent collection of essays on *Shakespeare and the Supernatural*, ed. Bladen and Brailowsky, but the category "supernatural" captures only a narrow band of a broader spectrum or religious ideas and influences.

19 See Richard G. Olson, *Science and Religion, 1450–1900*, 7–11.

20 The phrase comes from Descartes's *Treatise on Light* (*c.* 1629–33), also known as "The World." See *"The World" and Other Writings*, 32. Others translate the original French phrase as "laws that God has put into nature." See, e.g., Brooke, *Science and Religion*, 26.

21 For this point about Newton, see Brooke, *Science and Religion*, 16.

22 Bacon, *Great Instauration* (in *Works*, 4:20-21). Quoted in Harrison, *The Fall of Man and the Foundations of Science*, 4.

23 Harrison, *The Fall of Man and the Foundations of Science*, 6; Picciotto, *Labors of Innocence*, 9.

24 Picciotto, *Labors of Innocence*, 10, 13.

25 Ibid., 404.

26 Sarkar, *Possible Knowledge*, 174.

27 Killeen, *Unknowable*, 2.

28 Ibid., 3.

29 Topsell's *History of Four-Footed Beasts* mostly translates the *Historia Animalium* of the Swiss naturalist and philologist Gesner. See Todd Borlik's description of the work in *Literature and Nature in the English Renaissance*, 113.

30 Crane, *Losing Touch with Nature*, 8–9.

31 Donne, *An Anatomy of the World: First Anniversary*, (lns. 205–213), in *The Complete English Poems*, 276.

32 Crane, *Losing Touch with Nature*, 9.

33 Marchitello, *The Machine in the Text*, Chapter 5.

34 Elizabeth Harvey and Timothy Harrison, *John Donne's Physics*, 109. On Donne's engagement with astronomical discoveries, see also William Empson, "Donne the Space Man."

35 Marchitello, *The Machine in the Text*, 138, 135.

36 Donne, *Devotions upon Emergent Occasions*, 58.

37 Marchitello, *The Machine in the Text*, 120.

38 Ibid., 136.

39 See *OED*, s.v., "eccentric (adj.)," def. 1.a, and s.v., "concentric (adj.)," def 1.a.

40 Grant, *Science and Religion, 400 B.C. to A.D. 1500*, esp. 165–224.

41 Harvey, *De motu cordis*, 8, with English translation in *The Anatomical Exercises*, xiii. Boyle also alludes to the "magnificent fabric of the universe" in *Some Considerations Touching the Usefulness of Experimental Natural Philosophy*, 3.
42 Harrison, "The 'Book of Nature' and Early Modern Science."
43 Bodin, *Theatrum Universae Naturae*. For an important analysis of the work in its original context, see Ann Blair, *The Theater of Nature*.
44 One exception to this general tendency is Steven Matthews, *Theology and Science in the Thought of Francis Bacon*.
45 Bacon, *The Advancement of Learning*, ed. Kiernan, 24.
46 On the emergence of physico-theology out of Bacon, see Katherine Calloway, *Literature and Natural Theology in Early Modern England*, 9, and *Physico-Theology*, ed. Blair and von Greyerz. On eighteenth-century developments, see also Powell, *Perception and Analogy*.
47 Powell, *Perception and Analogy*, 14.
48 See Calloway, *Literature and Natural Theology in Early Modern England*, 38–9.
49 See Brooke, *Science and Religion*, 29.
50 Bacon, *Novum Organum*, ed. Jardine and Silverthorne, 82. See also William Eamon, *Science and the Secrets of Nature*, 9.
51 For the full range of historical meanings, see *OED*, s.v. "experiment (n.)."
52 Bacon, *The Advancement of Learning*, ed. Kiernan, 109.
53 Ibid., 64–5.
54 Puttenham, *The Art of English Poesy*, 99. In addition to Cowan's contribution to this volume, see her essay on "The Imagination's Arts: Poetry and Natural Philosophy in Bacon and Shakespeare."
55 See Greenblatt, *Hamlet in Purgatory*; Lake, *Hamlet's Choice*; and Lewis, *Hamlet and the Vision of Darkness*.
56 Sherman, *Second Death*, 165.
57 Swann, *The Poesy of Scientia in Early Modern England*, 8–9.

Part One

Divine Matter: Embodiment in Early Modern Science

1

Vesalius, Anatomy, and Salvation History

Sarah S. Keleher

Should we take Andreas Vesalius seriously when, in the midst of his landmark anatomy treatise *De humani corporis fabrica* (On the Fabric of the Human Body, 1543), he discusses his longing for his resurrection body? What about when he mentions God's creation of Adam's tongue?[1] Such moments are easy to dismiss as token gestures toward Christianity in an otherwise secular medical pursuit, particularly given that Vesalius played a pioneering role in anchoring anatomy in dissection in a way that anticipates modern medicine.[2] Because modern, Western science typically defines itself as secular, we may intuitively map Vesalius's modernization of anatomy onto the secularization of science. Indeed, scholars have long done so, especially in the wake of the nineteenth-century "conflict thesis," which depicted Christianity as enduringly hostile to scientific progress.[3] Although many historians have disproven the conflict thesis and it has fallen into disgrace in the history of science, it sometimes persists even in scholarship that seeks to complicate it.[4]

This essay pushes back against the lingering alignment of Vesalius with the secularization of anatomy. By investigating explicitly religious moments in the *Fabrica*, I uncover an alternative framework for the relationship between science and religion in early modern anatomy, identifying Vesalian anatomy as both a theological and scientific endeavor. I argue that the model of the body that Vesalius constructs in the *Fabrica* is as religious as it is scientific, for it is a soteriological model that depicts the body as a site of salvation history. Through his anatomical descriptions and the iconography of his illustrations, Vesalius embeds anatomy in a narrative of embodied salvation history that begins with God's creation of Adam and stretches through the Fall and Incarnation toward the promise of corporeal resurrection on the Last Day. I suggest that soteriological anatomy presents Vesalius with an intractable knowledge problem because dissection cannot fully satisfy the epistemological impulses generated by a soteriological understanding of the body. The prelapsarian body, the resurrection body, and the soul—all vital components of the soteriological body—are inaccessible to the anatomist's knife. Vesalius traces that knowledge problem back to the Fall.

He transforms the resurrection body into a deferred solution to the problem of fallen knowledge, merging epistemological, physical, and spiritual redemption in the resurrection body. He identifies the resurrection body as the endpoint of anatomy: the point at which the perfection of the body will render perfect anatomical knowledge accessible to the anatomist. In doing so, Vesalius engages in an act of theological knowledge production inseparable from his pursuit of anatomical knowledge. This vital presence of theological knowledge in the *Fabrica* demonstrates how a soteriological model of the body—a salvational model that underpinned early modern anatomy far beyond the *Fabrica*—bound together science and theology such that early modern anatomy could function as a site and source of religious understandings of the body.

A fundamental premise of my argument is that we should take Vesalius at his word when he discusses anatomy in relation to theology. Vesalius explicitly identifies anatomy as a religiously oriented act that seeks both to advance medicine and to uncover and glorify God's works in the human body.[5] He asserts that the study of anatomy demonstrates "the wisdom of the infinite Creator of the world" and links the body's importance to its status as "the lodging-house and instrument of the immortal soul" (lvi–lvii). Records indicate that both Catholic and Protestant religious professionals owned, read, and annotated the 1543 and 1555 editions of the *Fabrica*.[6] Protestant divine Philip Melanchthon also integrated material from the *Fabrica* into the revised *Liber de anima* (1552), while Catholic anatomy professors at the University of Ingolstadt absorbed Vesalian anatomy into the tradition of "Catholic medicine" that stretched back through Galen and Hippocrates to God's gift of medicine to Adam and Eve.[7] Although the *Fabrica* was dedicated to a Catholic monarch, owned by Catholic theologians, and used in Catholic anatomical education, it was also the target of Catholic censorship. The *Fabrica* earned a place on the *Index Librorum Prohibitorum* by virtue of its Protestant printer and its depictions of sexual organs.[8] While the treatise was neither burned nor fully banned, it was expurgated in Catholic countries. Those expurgations ranged from excising the name of the printer to blacking out the genitalia of anatomical figures. In one copy, the expurgations removed references to a handful of theologians in Vesalius's discussion of the soul (8:163-65).[9] Even when censors dedicated themselves to drawing little black shorts on the *Fabrica*'s nudes rather than on expurgating Vesalius's discussions of the soul, Catholic censorship marked the *Fabrica* as a text with tangible theological stakes.[10] Religious interest in the *Fabrica* was sufficiently strong that, at the close of the early modern period, church libraries owned more copies of the text than medical practitioners did.[11] Why should we take the religious aspects of the *Fabrica* seriously? Because many early modern readers did.

If we turn to the *Fabrica* with an eye for religion, we find that Vesalius's textual anatomization of the human form is shot through with a theological history of the body from its creation to its resurrection. God and his alternate manifestation, Nature, are active throughout the *Fabrica*.[12] They cover the brain with the trusty helmet of the skull (1:60), pad the buttocks with fat "to serve as a cushion" (1:144), arrange arteries (3:122), convolute nerves (4:198), shape the wrapping of the heart (6:65), and masterfully design the penis—an organ for which, according to Vesalius, "we should shower praise on the Creator of the world" (5:163). The *Fabrica* portrays God's creation of the body as an intricate, intentional process. When Vesalius explains the hard and earthy nature of bone, for example, he does so by stating that "God, the great Creator of all things, formed its substance to this specification with good reason, intending it to be like a foundation for the whole body" (1:1). He later discusses God's creation of Adam's tongue and his extraction of Adam's rib, embedding anatomy in narratives of divine creation (1:209, 2:196). Genesis provides the overarching blueprint of the *Fabrica*, whose division into seven books echoes the seven-day structure of God's creation of the world. Meanwhile, the *Fabrica*'s shorter companion volume, *De humani corporis fabrica librorum epitome* (known as the *Epitome*, 1543), illustrates "The Names of the External Regions or the Parts of the Human Body Visible Without Dissection" with nude figures of Adam and Eve.[13] The figures are fully intact rather than dissected, and they lack the keying mechanism through which Vesalius typically identifies anatomical features. As medical diagrams, they are impractical. However, that is not their primary function. Instead, their core function is to situate anatomical division in relation to salvation history and bodily integrity, affirming the continuity between the dissected body and the bodies of Adam and Eve. As a pivot point between the original form of the body created by God and the corrupted form of the body after the Fall, Adam and Eve capture one of the central tensions in early modern anatomy: the tension between anatomy as the glorification of God and the reality of anatomy as the dissection of fallen bodies. The Fall was simultaneously a limiting factor in dissection's ability to uncover knowledge of the prelapsarian body and a precondition for dissection, which depended on the mortality of the cadaver and, in the case of criminal cadavers, on the moral and spiritual degeneration that led to execution.[14]

Vesalius's dependence on cadavers limited his access to soteriological anatomy in another way as well. The ensouled nature of the body was crucial to its soteriological significance; however, dissection was confined to dead bodies whose souls had already departed. The *Fabrica* is deeply invested in the soul, as are many other early modern anatomy texts.[15] Vesalius discusses the workings of the soul in relation to anatomical features such as the heart (6:92-93), the animal spirit (7:162), and the brain (7:198). The workings of the soul were both a theological concern and a practical one for early modern anatomists because the soul was believed to play an active role in physiology.[16] Vesalius contends that to understand and treat the body properly, physicians must think about the soul and its faculties, including "in which part of the animal each is located, and most of all, if our mind is capable of conceiving such a thing, what is the substance or essence of the soul" (6:93). Vesalius's statement separates the essence of the soul from medical knowledge only by the distance of an *if*: "*if* our mind is capable of conceiving such a thing." Yet the soul quickly slips out of the anatomist's reach. Vesalius laments that the anatomical inaccessibility of the soul hinders knowledge about the aspects of the body most closely linked to it—among them, the faculties, the pulse, and respiration—which we cannot fully understand because "we still know nothing about the essence of the soul" (6:23). In seeking knowledge of the soul in the cadaver, anatomists must follow the logic of the trace, hunting for something that has already fled and that is innately inaccessible to dissection, being indivisible and incorporeal.[17]

Vesalius's discussion of the ventricles of the brain incorporates the soul in central ways. Vesalius begins Book VII of the *Fabrica*, which focuses on cerebral anatomy, with a chapter entitled, "The Brain Was Constructed for the Soul-in-Chief, for Our Senses, and for Movement Dependent on Our Own Whim" (7:161). Moreover, Vesalius retroactively maps the final three books of the *Fabrica* onto the model of the hierarchized tripartite soul.[18] He explains that Book V has examined the organs of nutrition and generation, in which the concupiscible or natural soul resides; Book VI has focused on the heart and the organs of respiration, which house the irascible or vital soul; and Book VII has studied the brain and the organs of the senses, along with the *principis animae*, which is the sovereign or rational soul (7:161). Following the standard Galenic schema, Vesalius asserts that each of those souls is linked to a fluid "spirit" produced by the associated organs. The liver produces "natural spirit," the heart produces "vital spirit," and the brain "produces animal spirit, the cleanest and thinnest spirit of all." The brain employs some of the animal spirit "for the divine workings of the soul-in-chief" and sends the rest "to the organs of sense and movement through the nerves as through cords" (7:161-62). Vesalius locates the production of the animal spirit in specific areas of the brain, where air and vital spirit "are transformed by the power inherent in the characteristic cerebral substance into animal spirit" (7:198). He is unwilling to go any further in speculating about the ventricular location of the rational soul and its faculties, however, stating, "I have no hesitation at all in saying that the ventricles produce animal spirit; but I cannot say anything about the areas in the brain occupied by the faculties of the soul in chief (though people who nowadays like to be called theologians and consequently think that there are no limits to their powers allot these as well)" (7:198). Vesalius posits the limits of anatomical knowledge at the border between the animal spirit and the rational soul, which are so entwined in the anatomy and physiology of the brain that anatomists cannot untangle them through empirical observation.

When Vesalius condemns self-styled theologians who think themselves able to allot "the faculties of the soul in chief" to specific regions of the brain, he is referring to the longstanding medical-theological tradition of ventricular psychology, which assigned the faculties of the soul to particular cerebral ventricles. Ventricular psychology was a common medical theory in early modern Europe; nonetheless, Vesalius condemns it as "a figment of the imagination of people who had never seen the brilliance of our Creator in the fabric of the human body" (7:164-65).[19] For Vesalius, the problem with ventricular psychology is that there is no tangible proof of it in the dissected body. The anatomist cannot point to a specific place in the cadaver's brain and show that the rational soul or one of its faculties resides there.[20] Thus, those who assign the faculties of the soul to areas of the brain claim theological-anatomical knowledge that is not justified by the body itself. In doing so, they uproot theology from the reality of the body. Vesalius's rejection of theological presumption with regard to the body and the body's relationship to the soul brings together theology and anatomy even as it seems to divide the two. By distinguishing dissection-based anatomical knowledge from the imaginings of foolish theologians, Vesalius suggests that dissection is the proper route to theologically valid knowledge of the body, even if that knowledge is limited in scope. Vesalius derides theologians who disregard the realities of anatomy not because he considers anatomy and theology incompatible but because he sees empirically based anatomy as the necessary grounds for a theological understanding of the body.

The dissection-based theological understanding of the body that Vesalius promotes in the *Fabrica* involves an awareness of the limits imposed on theological-anatomical knowledge by the condition of postlapsarian embodiment. Those limits come to the fore in his discussion of the theological problems posed by the similarity between human and animal brains. According to Vesalius, contemporary theologians categorically assert that the faculties of the soul are unique to human beings, but that assertion encounters a serious challenge in the reality of the body revealed by dissection. Vesalius remarks,

> All the theologians of our own time flatly deny that apes, dogs, horses, sheep, cattle and similar animals have the principal powers of the soul-in-chief and state that humans alone have [...] the faculty of reason and that (if I understand them correctly) all humans have this to an equal degree. Yet our dissections reveal that animals have the same number of ventricles as humans, and not only is the number the same but they are alike in every other respect as well except size and the integrity and accuracy of their temperament.
>
> (7:198-99)

If reason and "the principal powers of the soul-in-chief" reside in specific anatomical features, then there is little physical justification for denying those faculties to animals whose cerebral morphology is virtually identical to that of humans. Do animals, then, possess the faculties of the rational soul? By extension, do they *have* a rational soul? Although that theological problem is exacerbated by ventricular psychology, it is not exclusive to ventricular psychology. The problem is the inevitable product of locating the rational soul and its faculties in the brain, even if the anatomist refuses, as Vesalius does, to specify the precise location of those faculties. Anatomical inquiry threatens to reveal the lack of human exceptionalism, either reducing humans to the level of beasts or elevating animals to the status of reasoning, ensouled creatures.

Instead of offering a solution to the theological problem created by the similarity between human and animal brains, Vesalius ostentatiously turns away from the issue. He declares, "Even for the sake of humans, therefore, let me hold aloof from inquiring any further into the function of the ventricles" (7:199). Vesalius relocates the divinely wrought dignity of humankind from the anatomy of reason to the Christian soul and the resurrection body, inviting his readers to "sing hymns of praise to God the creator of all things and render thanks that he has given us a rational soul, which we share with the angels" (7:199). One "hymn of praise" in particular, Psalm 8, provides the intertextual scaffolding for this realignment of humans with angels: a transition from the brute reality of the dissected brain to the spiritual and physical dignity that God has granted to humankind. Psalm 8 voices two of the animating questions of early modern anatomy, asking, "What is man, that thou art mindful of him? and the son of man, that thou visitest him?" (8:4).[21] In the context of early modern anatomy, these questions function as ontological inquiries demanding an answer, propelling anatomical writing into the body as a site of salvation history. Psalm 8 weaves through early modern anatomy texts well beyond the *Fabrica*. Andreas Laurentius quotes from Psalm 8 in *L'Histoire anatomique* (The History of Anatomy, 1610), as does Helkiah Crooke in *Mikrokosmographia* (1615), and Johann Remmelin prints a portion of it alongside anatomical figures depicted as Adam and Eve in his anatomical flap book *Catoptrum microcosmicum* (Mirror of the Microcosm, 1619).[22]

Psalm 8 responds to those questions by turning to God's creation of humankind, proclaiming, "thou hast made him a little lower than the angels, and hast crowned him with glory and honour" (Ps. 8:5). Humans are intermediate creatures, of the earth but above it. The human body appears twice in the psalm: first, through "the mouths of babes and sucklings" (Ps. 8:4); and second, through the "feet" of man, under which God has placed all his works (Ps. 8:6). The body of God appears twice as well, each time in reference to his works: "the work of thy fingers" and "the works of thy hands" (Ps. 8:3, 8:6). God's presence lingers like a touch-memory in his creations. Bracketed by divine hands and human feet, the body evoked by the psalm blurs God and man. The intimacy with the divine that the psalm imagines is physical as well as spiritual, compacting the two forms of intimacy into divine presence.

In Psalm 8's account of the Creation, the answer to the question "What is man, that thou art mindful of him?" unfolds one life sphere at a time: "Thou madest him to have dominion over the works of thy hands; thou hast put all things under his feet; All sheep and oxen, yea, and the beasts of the field; The fowl of the air, and the fish of the sea" (8:5–8). The psalm reorders the sequence of creation in Genesis to proceed hierarchically, beginning with the heavenly bodies before progressing to humans and finally to the creatures over whom God has granted man dominion. The creatures themselves are presented in hierarchical order rather than in the chronology of creation. The reordered sequence locates man alongside the heavens and the angels in the space of the psalm, poetically enacting spiritual proximity and placing humans above all other earthly creatures in the unfolding time of the verse. The psalm's definition of man's creaturely dominion and spiritual elevation is implicitly prelapsarian, temporally bounded by the Fall in the overarching framework of the Genesis narrative. Psalm 8 draws its barriers before the moment of the Fall, redeeming man by sealing him in a stage of salvation history when redemption was not yet necessary, namely, the interval between the Creation and the Fall. That temporal seal is far from sound, however. The cycle of generation, of bodies that are born and die, seeps into the psalm through the phrase "the son of man," a reference that links humans' dual nature to Christ—the Son of God and son of man—when read in relation to the New Testament. Christ, whose redeeming sacrifice presupposes the Fall, provides not only an analogue for unfallen man but also a reminder of the fallen status of humanity and a promise of corporeal resurrection.

For Vesalius, Psalm 8 functions as a hinge between creation and resurrection. Vesalius shifts from contemplating the "rational soul, which we share with the angels" to offering thanks that

> by means of this soul, provided that faith is present as well, we may enjoy eternal happiness; for then our researches into the seat and substance of the soul will not have to be made by the dissection of bodies or by a faculty of reason that is crushed by the impediment of the body. He who is true wisdom will then be our teacher, for our existence will no longer be by means of this body that comes into being and then rots away but by means of a spiritual body very like his.[23]
>
> (7:199)

Vesalius's praise of God presents a salvation history of anatomy in which bodily resurrection promises epistemological redemption. The fallen body poses an impediment to knowledge about the embodied soul that will be removed through the purifying process of resurrection. Vesalius also attributes a perverse opacity to the fallen body. It is a fleshly "fetter" (*vinculum*) that "oppresses" (*obruta*) reason with claustrophobic corporeality. *Obruta*, the term that Vesalius uses to describe the body's oppression of reason, is as physical as metaphysical. It is a word that describes not only weighing down or oppression in the abstract sense but also the interment of a corpse.[24] The postlapsarian anatomist, whose rational soul is trapped by fallen flesh, can only approximate redeemed knowledge of the "seat and substance of the soul" through the mediation of his own fallen body and that of the dissected cadaver. The "impediment" of the anatomist's fallen body mediates between his rational faculty and the object of his knowledge, while the dead flesh of the cadaver mediates between dissection and knowledge of the bodies that the cadaver stands in for, which include the prelapsarian body; the living, ensouled body; and the resurrection body. The postlapsarian anatomist must dig for knowledge of the soul in the recesses of the dead and dissected body, with his search impeded by his own fallen form.

Vesalius suggests that salvation and resurrection will dissolve the epistemological barrier of the fallen flesh, enabling the resurrected anatomist to gain unmediated knowledge of the body and soul. The anatomist's re-embodied soul will finally be able to gain direct knowledge of itself in its embodied state. It is a mark of Vesalius's dedication to soteriological anatomy that unmediated knowledge of the anatomy of the soul is central to his vision of "eternal happiness." Vesalius's concept of a redeemed anatomical epistemology that does away with any mediation between the embodied anatomist and knowledge of the soteriological body is the apotheosis of the epistemological drive that led him to anchor anatomical knowledge in dissection. By privileging dissection and asserting the primacy of the body as the site of anatomical knowledge, Vesalius sought to minimize the mediation between the anatomist and the object of anatomical knowledge. Nonetheless, dissection was still a mediated practice, lacking direct access to the true epistemological object of dissection, which was not the cadaver *per se* but the human body in the fullness of salvation history, including the soul that dwelt within it. The epistemological constraints of the anatomist's fallen body, which prevent direct knowledge of soteriological anatomy by other means, render dissection necessary in the first place. But the resurrection body will no longer be bound by those constraints. The knowledge of body and soul accessible to the resurrected anatomist would still be mediated to some extent, for it would be received through teaching, but the teacher would be Christ. In the original Latin syntax, divine wisdom enwraps the anatomist's spiritual body as Christ imparts that knowledge. Rendered in English with the Latin syntax preserved, the quotation runs, "for he who is true wisdom, us no longer existing by means of this body that comes into being and then rots away but by means of a spiritual body very like his, shall instruct." The divine mediation of knowledge, couched as a pedagogical process, promises to lead to apprehension of truth. Mediation becomes revelation.

Vesalius asserts the continuity between earthly anatomical practice and divinely taught knowledge of the body even as he opposes the two. The research into the embodied soul that the resurrected anatomist pursues under the tutelage of Christ is a direct extension of the research that the anatomist has undertaken on earth "by the dissection of bodies." The divine anatomical teaching that awaits the resurrected anatomist is imagined as a purified form of earthly anatomical pedagogy: a divine anatomy demonstration in which Christ opens the body to his students' understanding without any taint of death. Resurrection enables the final fulfillment of the "researches" of dissection-based anatomy through an anatomical practice that is both an extension of the earthly anatomy demonstration and a transcendence of dissection, which has been purified out of existence by the shift from the fallen body to the resurrection body. The continuity between earthly and heavenly anatomical practice is rooted in the soteriological and anatomical continuity of the body itself, for which Christ's body serves as a guarantor. Christ's body—incarnate, crucified, resurrected, and glorified—is the copy text for the salvation history that binds all Christian bodies.

Vesalius anchors his depiction of the post-resurrection divine anatomy lesson in the "spiritual body" that the anatomist will then possess, which he describes as "very like" (*quàm similimo*) that of Christ. The body is a common tie that binds humans to Christ in resurrection just as it did in the Incarnation. Whereas Christ took on the vulnerable flesh of postlapsarian humans in the Incarnation, humans take on "spiritual bodies" like that of the glorified Christ in the general resurrection. Resurrection involves a change in bodily nature but not an abandonment of the body, which remains central to humans' soteriological significance in relation to the Son of God. The contrast that Vesalius draws between the mortal body and the spiritual body ultimately derives from St. Paul's description of corporeal resurrection in 1 Corinthians 15. Paul describes the mortal body as a seed that will sprout into new growth at the resurrection through the power of God. That new growth will be a "spiritual body" (1 Cor. 15:44). The body, Paul states,

> is sown in corruption; it is raised in incorruption: It is sown in dishonor; it is raised in glory: it is sown a natural body; it is raised a spiritual body. There is a natural body, and there is a spiritual body [...]. The first man is of the earth, earthy: the second man is the Lord from heaven [...]. And as we have borne the image of the earthy, we shall also bear the image of the heavenly.
>
> (1 Cor. 15:42-49)

Paul's description of the earthly body might as well be a description of the criminal cadaver, characterized by its corruption, dishonor, and earthly materiality. The spiritual body, incorruptible and glorified, is both diametrically opposed to the earthly body and continuous with it, raised up from it as from a seed.

By the early modern period, the dominant interpretation of the corporeal resurrection among Protestants and Catholics alike was the doctrine of the resurrection of the same body, which held that people would arise at the resurrection in the same bodies that they possessed in life.[25] The most influential authority for such a position was St. Augustine, who argues that the resurrection body retains the same material as the earthly body, albeit "clothed with incorruption and immortality."[26] It will be a "spiritual body" not because it will consist purely of spirit but because the flesh will serve the spirit with perfect obedience. One of the glorious aspects of the resurrection body, Augustine suggests, is that its organs will be open to view, perhaps through the mechanisms of heavenly sight or perhaps because the resurrection body will be transparent. The "corporeal harmony" of the body, expressed in the "beauty of proportion in the inner organs" as well as in the exterior parts, "will appear arranged through all parts of the body, within and without," and "will inflame the hearts of all rational beings to praise their great Creator."[27] The body reveals itself through an incision-less dissection. It is the same body that it was on earth, reassembled from the same fragments and reunited with the same soul. Simultaneously opposed to the fallen body and linked to it through salvation history, the spiritual body promises the triumph of the body over mortality and the triumph of anatomical knowledge over the constraints of the flesh. No wonder Vesalius yearns for it.

Vesalius grafts the history of anatomy onto embodied salvation history by identifying the general resurrection as the point when anatomical inquiry, like the body itself, will be brought to a purified fruition. By grafting anatomical practice onto salvation history, Vesalius asserts the compatibility of dissection-based anatomy and resurrection. Embedded in such an assertion of compatibility is a theological claim about the relationship between the earthly body and the resurrection body. Vesalius suggests that there is sufficient continuity between earthly and resurrection bodies for anatomy to be meaningfully the same endeavor before and after the resurrection and for the seat of the soul to be the same in both iterations of the body. The doctrine of the resurrection of the same body deepens that anatomical continuity by coupling it with material continuity.

Unfortunately, the continuity between the cadaver and the resurrection body could also pose a theological problem for anatomy. If resurrection requires the reconstruction of the body from the same anatomical components that composed it during life, could dismantling a cadaver during dissection prevent its resurrection? What if dissection destroyed some aspect of the body necessary for its resurrection? Similar anxieties persisted well into the nineteenth century, when *The Boston Medical and Surgical Journal* published an article entitled "Dissection Viewed with Reference to the Resurrection" that attempted to defend dissection against claims that it prevented resurrection.[28] In the *Fabrica*, Vesalius defends against such anxieties through his refutation of the theory that the resurrection body springs from a specific bone—in this case, one of the sesamoid bones, which are small, rounded ossicles located in the tendons near the joints of the thumb and big toe. Vesalius states that "magicians and followers of the occult" claim that the second ossicle at the internode of the big toe is "a bone [...] which resists all decay, which remains hidden like a seed in the ground after death, and which will reproduce the person at the last day of judgement." The resurrection ossicle allows for an anatomically precise interpretation of the Pauline seed, drawing together Paul's description of the resurrection and Ezekiel's vision of God reanimating the dead from dry bones, discussed below (Ezek. 37:1-10). As one of the sesamoid bones, the ossicle is linked to the Pauline seed at the level of language and appearance (the term "sesamoid" derives from the ossicle's resemblance to a sesame seed). Vesalius rebuffs the belief that the ossicle is the resurrection bone by demonstrating that it is vulnerable to destruction. He states that "the bones themselves do not support the Arabian fabrications, for they can in fact be broken and burnt just like other bones," even if they are "surprisingly resistant to decay" (1:298).[29] Since the supposed indestructibility of the ossicle is the grounds for identifying the ossicle as the resurrection bone, the anatomist's destruction of it disproves the claim that it is instrumental to resurrection. Underlying Vesalius's rejection of the theory of the resurrection ossicle is a rejection of the idea that anatomists might destroy the possibility of a person's resurrection by dismantling their body. Vesalius suggests that the body is inherently vulnerable to destruction and that we cannot account for resurrection by anchoring it in postmortem bodily integrity.

In practice, dissection functioned as the inverse of resurrection. If we turn to Ezekiel 37, one of the key Biblical passages for the corporeal resurrection, we find God reanimating the dry bones of the dead by rebuilding the body layer by layer. Ezekiel reports that, through God's power, "the bones came together, bone to his bone. And when I beheld, lo, the sinews and the flesh came up upon them, and the skin covered them above [...] and the breath came into them, and they lived, and stood upon their feet" (Ezek. 37:7-8, 10). God rearticulates the skeleton, stretches the sinews back over it, covers it with flesh, seals it with skin, and reanimates it with living breath. Dissection does the opposite with equal meticulousness. The anatomist uncovers the body layer by layer, peeling back skin, cutting away flesh, slicing sinews, and paring the body down to its bones. In early modern anatomy demonstrations, the dismantling of the body was systematic and performative. The anatomist divided the body into parts both conceptually and physically, through a combination of brute force and minute delicacy.[30] Over the course of a few days, anatomy demonstrations enacted the dismantling of the body that churned ceaselessly onward in the hidden recesses of the grave. Such dismantling might continue even after dissection if the anatomist decided to preserve the cadaver's bones for further study, a process that involved scraping the bones clean and boiling them.[31] By the end of that process, the anatomist would have fully reversed the trajectory of God's rebuilding of the body from dry bones. Still, the raw materials of resurrection would remain solely within God's power to reassemble. Dissection would remain part of a larger soteriological history that stretched beyond death toward life.

Even as they deconstructed bodies during dissection, anatomists could also prefigure God's reconstruction of the body by textually rearticulating cadavers into coherent models of the body in anatomy treatises and by materially reconstructing cadavers as articulated skeletons, quite literally bringing "bone to his bone." Vesalius's instructions in the *Fabrica* on how to construct an articulated skeleton describe the process of articulation as akin to "repairing broken plates and assembling stone cauldrons," literalizing Tertullian's description of the reconstruction of the body during resurrection as the mending of a broken pot (1:382).[32] Vesalius's definition of the term "skeleton" as "the whole integrated structure of the mutually interconnected bones" underscores the symbolic value of the articulated skeleton as an emblem of restored bodily integrity (1:29). Anatomy was so closely associated with such skeletons in early modern England that "anatomy" was a term for a skeleton as well as for the overall discipline of anatomy, the practice of dissection, and the body that came under the anatomist's knife.[33] In early modern anatomy, articulated skeletons served as pedagogical tools for teaching bone morphology and as anatomical models at dissections, where they could provide the audience with a skeletal scaffolding on which to map the features of the opened cadaver. Vesalius instructs his readers to have an articulated skeleton at hand during dissections, and he himself used an articulated skeleton to show his audience the position of the bones during anatomy demonstrations (5:212-14).[34] A skilled anatomist could construct an articulated skeleton from the remains of a dissected cadaver—and the same body that had been dismantled in one dissection could reappear in reconstructed form at another as an articulated skeleton, freed from fleshly decay. Reassembled and cleansed of putrefaction, the articulated skeleton presented an image of mortality swept up in the first stirrings of resurrection. It reminded viewers that dissection was merely one stage in an ongoing soteriological narrative that preserved the body's transtemporal integrity.[35]

The use of iconographically significant supports to fix articulated skeletons upright intensified their symbolic function in anatomy theaters and anatomy treatises. In his instructions for constructing an articulated skeleton, Vesalius directs his readers to pose the skeleton upright and prop its arms up by affixing its hands to a stake in the form of "a scythe or some other object" (1:380-82). In the articulated skeletons depicted in the *Fabrica*'s frontispiece and osteological plates, the stakes take the form of a scythe and a spade, combining practical support with a Christianized *memento mori* (frontispiece, 1:387). Such skeletons were a material part of anatomy demonstrations, iconographic supports and all. Even the emotive position of the skeleton on the *Fabrica*'s frontispiece, with its head tilted back and its jaw open as if in a lamenting cry, corresponds to one of the options that Vesalius gives for positioning the skull and jawbone in his instructions for crafting an articulated skeleton (1:382). The pedagogical function of those skeletons was religious as well as morphological. In the *Fabrica*'s frontispiece, the scythe-wielding articulated skeleton looms over the partially dissected female cadaver on the anatomy table while Vesalius opens the cadaver's abdomen to expose her womb. As a *memento mori*, the articulated skeleton demonstrates the human subjection to death initiated by the Fall. The skeleton is aligned with the pregnant cadaver's opened womb such that birth becomes an entry into death. Nonetheless, the upwards visual movement from the supine cadaver to the upright skeleton also traces a trajectory of the dead rising. The articulated skeleton merges the lingering animation of the newly dead cadaver with the resurrective reanimation that awaits her when stripped down to the bone.[36] The skeleton is transformed from a symbol of mortality *tout court* into a symbol of the promise of resurrection implicit in death.

Vesalius builds resurrective reintegration into dissection at the level of the text as well through the sequence of the *Fabrica*. He begins the *Fabrica* with the anatomy of the bones, then moves to the surface of the body, reconstituting the body around the skeleton to anatomize skin, flesh, and muscles. From there, he continues into the body's depths, describing the anatomy of the veins, nerves, and organs. Vesalius concludes the *Fabrica* with a chapter on animal vivisection, reanimating the body to investigate the living form before death claims it once more. The *Fabrica*'s resurrective gathering of parts continues when its pages are bound. Textual binding was part of dissection itself; when excising organs, Vesalius ligated the surrounding vessels with bookbinding thread to prevent the outflow of fluids (5:224). The ligated body would be dismantled as dissection progressed, but the anatomy text would continue to gather together the body's scattered parts.

Reconstructing and reanimating the dismantled body, the *Fabrica* renders anatomical writing a resurrective art. As Psalm 8 asks, "what is man, that thou art mindful of him?" In the pages of the *Fabrica*, man emerges as an intricate structure, embodied and animate, bound together by the bookbinder's art and—most importantly—by salvation history, even when physically dismantled. Reconceptualizing the *Fabrica* as a text deeply invested in soteriological anatomy allows us to see a more productive role for religion in early modern anatomy: a role in which Christianity expanded the imaginative, representational, and epistemological possibilities of Vesalian anatomy rather than impeding anatomical inquiry.

Notes

1 See Vesalius, *On the Fabric of the Human Body*, trans. Richardson and Carman, 7:199, 2:196, translated from Vesalius, *De humani corporis fabrica* (1543). All quotations from this work of Vesalius are to this edition, which

is hereafter cited parenthetically by book and page number. All citations that explicitly refer to or quote from the original Latin of the *Fabrica* refer to Vesalius, *De humani corporis fabrica*, 1543, with corresponding book and page numbers.

2 Vesalius modernized anatomy in multiple ways, anchoring anatomical knowledge in dissection, representing the body in minute anatomical and physiological detail, and supplementing detailed anatomical descriptions with keyed illustrations. For further information on Vesalius's foundational role in early modern and modern anatomy, see Andrew Cunningham, *The Anatomical Renaissance*, 131; Valeria Finucci, "Vesalius and the Languages of Anatomy," 2; Devon Hodges, *Renaissance Fictions of Anatomy*, 2; Dániel Margócsy, Mark Somos, and Stephen M. Joffe, "The Art of Medicine," 738; and Gary Steiner, "The Cultural Significance," 275. For the influence of the *Fabrica* and the spread of Vesalian anatomy in England, see Margócsy, Somos, and Joffe, "The Art of Medicine," 738; Margócsy, Somos, and Joffe, *A Worldwide Descriptive Census*, 5, 131; Jonathan Sawday, *The Body Emblazoned*, 56–7; and Richard Sugg, *Murder after Death*, 2.

3 For a foundational example of the conflict thesis, see Andrew Dickson White's *A History of the Warfare of Science with Theology in Christendom*. White asserts that Vesalius fought against Church oppression as a "champion of scientific truth" to advance dissection-based anatomy, which emerged "in spite of theological dogma" (50).

4 For an overview of the rejection of the conflict thesis in history of science, see James C. Ungureanu, "Relocating the Conflict," 1106–30. For rebuttals of the conflict thesis with regard to medieval and early modern anatomy, see Cunningham, *The Anatomical Renaissance*; Katharine Park, *Secrets of Women*. For an example of the subtle persistence of the conflict thesis in scholarship that seeks to complicate it, see David Hillman's depiction of the relationship between Christianity and anatomical science in early modern concepts of "visceral knowledge." Hillman notes in "Visceral Knowledge" that early modern Christianity established the importance of the bodily interior as "the ultimate site of faith," observes that "virtue and sin [...] were often imagined as literally inhabiting bodily innards," and perceptively asserts that "the ascription of omniscient knowledge of the human interior to an all-seeing god [...] renders human innards the central locus of connection with the divine." However, he goes on to suggest that the Christian mode of understanding the body as "the ontological site of belief" and the emergent mode of understanding the body as "the epistemological site of rapidly growing medical and anatomical knowledge" in the sixteenth century were "*incompatible* in terms of the kind of access to the body's interior they deem possible" and "jostled against one another" (82, 85–6, emphasis added). Vestiges of the conflict thesis also appear in Sawday's account of the "early modern culture of dissection," which has been foundational to the last thirty years of scholarship on early modern

anatomy—this essay included. In *The Body Emblazoned*, Sawday does emphasize that Christianity was core to the early modern paradigm of "sacred anatomy," which he describes as "an understanding of the body which sought not just to divide the corpse but to assert the centrality of the body, even in division, to the key articles of the Christian faith." Within that paradigm, "the body's construction was evidence for the work of God, and the doctrine of the incarnation endowed the body with a special status" (107). Sawday argues that "anatomists and the theologians were, for a time, able to work with one accord" (106), asserting that before the "release of the body from theology" in the eighteenth century, "anatomists struggled to express a harmonized view of the created fabric of the body, so that their own researches and scriptural doctrine, rather than jostling one another, simply replicated the authenticity of each system of knowledge" (126). While he acknowledges affinities between theology and anatomy, traces of the conflict thesis still appear in his occasional rhetoric of opposition and oppression. He asserts, for example, that "the history of the representation of the body in science in the early-modern period was a negotiation between [...] two *irreconcilable* positions": one in which the body remained situated "within a theologically ordained universe," and another in which "the body was *liberated from theology*, only to be made subject to the equally stringent demands of the scientific method" (98, emphasis added). Sawday positions Vesalius as a hinge point in that paradigm shift, suggesting that, although Vesalius "sought to remind his audience of the tradition of 'sacred anatomy' on the title-page of his own work," the program of "religious didacticism" through which pre-Vesalian depictions of the bodily interior sought to represent "the divine nexus which united the body, the soul, and God into a harmonious scheme" was "fractured with the advent of the post-Vesalian practice of anatomy" (105).

5 Vesalius is not alone in doing so. Other early modern anatomy texts that do so include Crooke, *Mikrokosmographia*, 12; Laurens (Laurentius), *L'Histoire anatomique*, 21; and Remmelin, *Catoptrum Microcosmicum*, sig. Br. For further discussion of early modern anatomy as a pursuit centered on demonstrating God's goodness in the human body, see Cunningham, *The Anatomical Renaissance*, 39; Cunningham, "Protestant Anatomy," 45; Hodges, *Renaissance Fictions of Anatomy*, 4; Peter Mitchell, *The Purple Island and Anatomy*, 123; Katherine Rowe, "God's Handy Worke," 288, 304–5; Sawday, *The Body Emblazoned*, 106–17, 130; and Steiner, "The Cultural Significance," 276–7.

6 Margócsy, Somos, and Joffe demonstrate that "The *Fabrica* was read and owned equally by catholics and protestants" and that although "Lutheran Wittenberg was clearly an important city for spreading Vesalius's fame, a similar emphasis on the religious and social implications of the *Fabrica* can also be observed in Calvinist and catholic circles." See *A Worldwide Descriptive Census*, 33, 112. See also Margócsy, Somos, and Joffe, "The Art of Medicine," 739.

7 Jürgen Helm, "Religion and Medicine," 54–67. For further discussion of Philip Melanchthon and Vesalian anatomy, see Davide Cellamare, "Anatomy and the Body," 144–51; Cunningham, *Anatomical Renaissance*, 230–4; Tricia M. Ross, "*Anthropologia*," 10; and Sugg, *Murder after Death*, 88.

8 Margócsy, Somos, and Joffe, *A Worldwide Descriptive Census*, 121. Vesalius's choice of the Protestant printer Johannes Oporinus is sufficiently odd for a Catholic anatomist that Cunningham has marshaled it as evidence that Vesalius may have been a Lutheran. Cunningham acknowledges, however, that among Vesalius's contemporaries, "no one remarked that Vesalius was not Catholic," and Vesalius's affiliation with Charles V, to whom he dedicated the *Fabrica*, seems to align him with Catholicism (*The Anatomical Renaissance*, 230–5, 224). Cunningham has since backtracked somewhat, asserting in "Protestant Anatomy" that "I am not claiming that Vesalius was a Protestant by religious persuasion: but I am claiming that his mode of approach to anatomizing, which was to be enormously significant in its own time and highly influential, was Protestant in its structure and characteristics" (47). The evidence is simply too thin to argue convincingly that Vesalius was anything but Catholic. What we *can* say with confidence is that he exercised a significant impact on thinking about the body among both Catholics and Protestants.

9 Margócsy, Somos, and Joffe, *A Worldwide Descriptive Census*, 121–30.

10 Yes, there is a copy of the *Fabrica* in which the censor has given two of the nude musclemen tiny shorts. See Ibid., 126.

11 Ibid., 739.

12 John Banister makes the identification between God and Nature explicit in the context of anatomical creation in his anatomical treatise *The History of Man*, which draws heavily on the *Fabrica*. The listing for "nature" in the index of *The History of Man* reads, "Nature, that is, God." Elsewhere in the treatise, Banister states, "neither was it the mighty pleasure of God (whom we call Nature) nor any parcel of his decree, that the substance of the bones at all should be made sensible" (sigs. Jiijr, Cjr).

13 Vesalius, *De humani corporis fabrica librorum epitome*, sigs. Kv, Lr. Translation by Lind and Asling in *The Epitome of Andreas Vesalius*, 90. The link to Genesis is a common feature of early modern anatomy texts. Remmelin, for example, begins the main text of his anatomical atlas *Catoptrum microcosmicum* with an account of the history of the body drawn from Genesis (sig. Bv).

14 Patristic and medieval theology cultivated a belief that the human body had changed in concrete, physical ways after the Fall, making the body subject to death and pathology. See Nancy G. Siraisi, *History, Medicine, and the Traditions of Renaissance Learning*, 28–29. Vesalius mentions the marks of execution on cadavers at multiple points in the *Fabrica* (2:174, 7:239). For further discussion of the use of criminal cadavers in early modern

anatomy to demonstrate the glories of God, see Hillary M. Nunn, *Staging Anatomies*, 64; Sawday, *The Body Emblazoned*, 63–4, 106–7. Sawday in *The Body Emblazoned* highlights that "the type of fallen humankind who, in the guise of an executed criminal corpse, was carried into the anatomy theatre, was also the potentially redeemed figure of the 'second Adam,'" and "this framework of belief" operated not only in early modern anatomy demonstrations but "at every level in Renaissance anatomical texts even when, as practical guides to dissection, they sought to instruct the trainee physician in his craft" (106).

15 See, for example, Crooke, *Mikrokosmographia*, 12; and Laurens, *L'Histoire anatomique*, 22. For further discussion of early modern anatomy and the soul, see Cunningham, *The Anatomical Renaissance*, 196; Mitchell, *The Purple Island and Anatomy*, 122; Steven Pender, "Essaying the Body," 235; Sawday, *The Body Emblazoned*, 16; and Sugg, *The Smoke of the Soul*.

16 Sugg, *The Smoke of the Soul*, 15, see also 73.

17 For the widespread early modern belief that the soul was an "incorporeal substance," see Sugg, *The Smoke of the Soul*, 1.

18 The tripartite model that Vesalius adopts follows that outlined by Galen. See *De usu partium* (On the Usefulness of the Parts of the Body) IV.13, 228–9.

19 Ventricular psychology appears, for example, in the medical works of Robert Burton and Thomas Willis. See Burton, *The Anatomy of Melancholy*, ed. Jackson, 154; and Willis, *Two Discourses*, 35–6.

20 Even René Descartes, who would later identify the pineal gland as the seat of the soul based on concrete features of its morphology, could not directly show the soul's presence in it. See *The Passions of the Soul*, I.31.

21 Biblical citations refer to the 1611 King James Version. While the Geneva Bible is closer in publication date to the *Fabrica*, the translations in the King James Version more closely approximate the vulgate in key areas for the *Fabrica*, such as Psalm 8. Accordingly, the King James Version enables a more accurate analysis of the relationship between the *Fabrica* and the Bible, as the vulgate is the version of the Bible that Vesalius would have known.

22 Laurens, *L'Histoire anatomique*, 2–3; Crooke, *Mikrokosmographia*, 3; and Remmelin, *Catoptrum microcosmicum*, sig. Er.

23 I have altered Richardson and Carman's translation in changing the phrase "a spiritual body exactly like his" to "a spiritual body *very* like his" in order to capture more closely the sense of *quàm similimo* in the original Latin. The full Latin version of the passage is as follows: "Ac proinde Deo rerum omnium opifici hymnos canentes, gratias agemus, quòd nobis rationalem animam, quam cum angelis [...] communem habemus, largitus fuerit, cuius beneficio si modò fides adsit, aeterna illa fœlicitate fruemur, quando animae sedem ac substantiam non corporum resectione, aut nostra corporeo vinculo obruta ratione inquirere necessum erit. Nam is qui vera

est sapientia, nos non hoc generabili & putrescibili corpore amplius, sed spirituali suoque quàm similimo constantes, edocebit" (7:636).

24 *A Latin Dictionary*, ed. Lewis and Short, s.v. "obruo."

25 Kevin Chovanec, "Stirring Meats," 5. See also Lloyd Strickland, "The Doctrine of 'The Resurrection of the Same Body,'" 168. For another recent study of these debates, see Daniel Juan Gil, *Fate of the Flesh*, esp. 4–10. Caroline Walker Bynum notes that "a concern for material and structural continuity" in theorizations of the resurrection body "showed remarkable persistence" throughout the Middle Ages, for "person was not person without body." Bynum, *The Resurrection of the Body*, 11.

26 Augustine, *City of God* 7.22.21, trans. Green, 301 and 4.13.20, trans. Levine, 213.

27 Augustine, *City of God* 7.22.24, trans. Green, 333 and 7.22.30, 373. See also Bynum, *The Resurrection of the Body*, 100, 140.

28 "Dissection Viewed with Reference Resurrection," 197–99.

29 The Latin verb that Richardson translates as "reproduce" is *produco*, which brings together a sense of production, generation, and cultivation particularly apt for the seed-like sesamoid bone in the context of St. Paul's description of the earthly body as the seed from which the spiritual body sprouts. *A Latin Dictionary*, ed. Lewis and Short, s.v. "produco." For the belief that the second ossicle of the big toe may be the resurrection bone, see also Crooke, *Mikrokosmographia*, 1110.

30 For the systematic dismantling of the body in dissection, see Rafael Mandressi, "Of the Eye and the Hand," 74.

31 See Vesalius's instructions for disassembling and cleaning a cadaver's bones prior to reconstructing them into an articulated skeleton (1:370-384).

32 See also Bynum, *The Resurrection of the Body*, 36.

33 *Oxford English Dictionary*, s.v. "anatomy."

34 For an eyewitness account of Vesalius's use of an articulated skeleton during an anatomy demonstration, see Baldasar Heseler, *Andreas Vesalius' First Public Anatomy*, 89.

35 For the use of a dissected cadaver as the source of the bones for an articulated skeleton, see Vesalius, *On the Fabric of the Human Body* (6:107-16). For further discussion of the articulated skeleton as a "kind of reanimated body," see Rose Marie San Juan, "The Turn of the Skull," 969.

36 Sawday similarly identifies the *Fabrica*'s frontispiece—along with the animate anatomical figures of the 'musclemen' and osteological plates—as examples of "sacred anatomy." Such images, Sawday contends, serve the need for "the body [...] to be contextualized, juxtaposed with other motifs, shown in extravagant postures, in order to demonstrate the full human and divine significance of the anatomist's skills." For Sawday, the "moral lesson of 'sacred anatomy'" in Vesalius's anatomical illustrations is primarily the body's "material fate at the hand of time" and "the triumph of death [...] over human affairs," although the animate anatomical

figures also "paradoxically [...] appeared to rejoice in a liberation from death guaranteed them by the reductive process of dissection" (*The Body Emblazoned*, 115–16). Nunn provides a helpful counterpoint to an emphasis on the triumph of death. According to Nunn, animation in early modern representations of cadavers reflects the lingering animation attributed to corpses in early modern English understandings of death, not the reanimation of a thoroughly dead corpse (*Staging Anatomies*, 73). While my interpretation is closer to Sawday's, I suggest that lingering animation and anatomical *re*animation converge in the articulated skeletons of the *Fabrica*. I argue that the core ideological lesson of the *Fabrica*'s frontispiece, as with its 'musclemen' and osteological plates, is the immanence of corporeal resurrection in the dead body rather than the inevitability of bodily decay and mortality.

2

Strange Evidence: Hymens Between Science and Religion

Margaret Ferguson

Duke Senior: *Dost thou believe, Orlando, that the boy*
Can do all this that he hath promised?
Orlando: *I sometimes do believe and sometimes do not.*
—Shakespeare, *As You Like It*, 5.4.1-3

Gower: *Hymen hath brought the bride to bed,*
Where by the loss of maidenhead
A babe is molded ...
—Shakespeare, *Pericles*, 3.0.9-11

Now faith is the ground of things, which are hoped for,
and the evidence of things, which are not seen.
—St. Paul, Letter to the Hebrews 11:1, Geneva Bible (1560)[1]

Introduction

The "boy" in whose power Shakespeare's Orlando "sometimes" believes and "sometimes" does not claims that his knowledge comes from a magician: "Believe then, if you please, that I can do strange things. I have since I was three years old conversed with a magician most profound in his art, and yet not damnable" (5.2.53-6). Emphasizing but also sowing doubts about the legitimacy of the source of his knowledge, Ganymede (as the transvestite Rosalind has named herself in a bold joke on homoerotic passion) promises to bring Orlando's beloved to him "tomorrow—human as she is and without any danger" (5.2.61).[2]

What might such danger be or mean? How is it yoked to a "not damnable" exercise of magic? The contours of an answer may appear in the play's final scene, where Rosalind/Ganymede produces an avatar named "Hymen" who claims to be able to "bar confusion" and bring "conclusion" to the play's many erotic and political plots (5.4.116-17). Hymen also promises to make "mirth in heaven"—a place he seems to know well—by effecting the "atonement" (at-onement) that occurs, Hymen says, "when earthly things" are "made even" (5.4.100). In the scene before Hymen's entrance, the character whom Orlando knows as Ganymede reiterates their magical plot-maker's vow: "I have promised to make all this matter even [...] / and from hence I go, / To make these doubts all even" (5.4.18, 24-25). Paradoxically, the even number of two is enjoined to become the odd number one in an enigmatic allusion to an "earthly" version of the central Christian doctrine of Atonement: Christ's sacrifice for the sake of fallen mankind.

Far from barring doubts, Shakespeare's Hymen generates and reshapes them. He does so in a politically and theologically risky way that raises questions about belief, faith, and their vexing relation to questions of evidence.[3] Such questions illustrate how different categories of belief cut across religious, literary, and scientific epistemologies, exposing tensions in and among the three major meanings—and the multiple interpretations—of the foreign word "H/hymen."[4] As the proper name of an enigmatic Greek god of marriage rites who was represented both as impotent and as cosmically potent;[5] as the name for a contested female body part (or site) that allegedly served as a sign of virginity; and as a name for the institution of procreative marriage itself, H/hymen (*Hymenaeus* in Latin) became, over centuries and across borders of geography, language, and religious belief, an instance of what M. M. Bakhtin calls a "chronotope of the threshold."[6] In analyzing a broad range of representations of both Hymen and the elusive hymen, this essay examines a species of doubt that stands at a significant nexus of science, literature, and religion. It typically recurs over time rather than proceeding toward a resolution. The kind of doubt that interests me here thus exists in a recursive rather than a linear time, one that extends into the present, when "hymenoplasty" (hymen-replacement) surgeries continue to be performed in many places around the world, although they have been declared illegal—medically unnecessary and conducive to "myths" about female virginity—in Great Britain.[7] And in Sweden, "hymen" as an anatomical term has been officially replaced by "vaginal corona," which carries its own mythical associations but is intellectually useful in that it queries the still-dominant image of the somatic hymen as a "seal" that the male penis must destroy in order to deposit its seed.[8] Major studies in the history of anatomy have simply avoided the hymen and the specters of shame and doubt it raises in conjunction with the figure of a hyper-potent or impotent pagan god often doubly invoked as Hymen/Hymenaeus in Latin texts.[9]

Zones of Doubt

The word "hymen" elicited disputes about belief from its first appearance in the Greek-influenced written record as a term denoting a specifically human female membrane. It occurs in a gynecological treatise by Soranus of Ephesus written in Greek for literate midwives in Rome around the turn of the second century CE. In previous scientific writings in Greek, as Giulia Sissa has shown, "hymen" signified a skin-like membrane that protected bones and organs in all warm-blooded animals.[10] But by the time Soranus chided his readers about their false belief in a "hymen" that sealed the human female's reproductive body, its meaning had changed, likely in conjunction with early Christian and Jewish debates about the somatic purity of Christ's mother Mary. A belief in a virginal seal, allegedly broken during the event we call "defloration," was a "*pseudos*," a superstitious error or counterfeit in Soranus's experienced view.[11] I will return later to his *Gynaikeia* (Gynecology, *c.* 98 138 CE), which distinguishes between an unnatural hymen—a rare and pathological phenomenon—and a natural state in which hymens imagined as seals over the vagina simply do not exist in all bodies born human and classified as female.

Servius, the influential fourth-century CE commentator on Virgil's *Aeneid*, shares Soranus's skepticism about the existence of a "virginal membrane" as a universal female attribute—what Sigmund Freud later reified as an "organ" (in "The Taboo of Virginity," 1918), as if the hymen were like a uterus or a clitoris.[12] Servius finds the idea that "hymen" signifies a "virginal membrane" to be barely credible and reports what "some say" (*dicitur*) about it with "wary circumspection."[13] In another gloss on Virgil's poem—on a line describing Helen's adulterous union with Paris as an "unlawful marriage" (*inconcessos* [...] *hymenaeos*; *The Aeneid*, 1:651), Servius judges between a comic and a tragic origin story for the marriage god's name. The tragic one envisions Hymenaeus as a "young man who on his wedding day was buried beneath the walls of his house." For that reason, Servius writes, "he is named in weddings by way of expiation."[14] But this ancient etymology of the wedding god's name "is false," Servius states. He thus dismisses an etymology of considerable somatic complexity: the marriage god who dies on his wedding night when (according to Pindar) his "skin is first touched in marriage" evokes not only the bride's loss of her maidenhood *(parthenia)* but also the groom's loss of his erection.[15] The truest story, most worthy of belief, Servius also suggests, has a happy ending and is a good omen for marriages: it is a tale about a poor but beautiful and heroic Athenian boy who disguised himself as a maiden to follow his beloved in a group of maidens journeying to worship at Eleusis.

Pirates kidnapped the maidens on the way home, but Hymen came to their rescue, killing their captors and negotiating the terms of a marriage with his beloved's father before returning to Athens and a fortunate wedding.

This strange figure of a mortal man deified by a happy marriage achieved both through strength and clever deceit (his transvestite disguise, his manipulation of his bride's father) appears in several under-appreciated texts that likely influenced Shakespeare's *As You Like It*. One is Vincenzo Cartari's encyclopedic and mythographic text about the ancient gods, which I would categorize as an exercise in philology and comparative religion (1556, rev. 1571). Another, in English, is George Chapman's long digression on Hymen, influenced by Cartari, in Sestiad V of Chapman's continuation of Marlowe's *Hero and Leander*. This was first published as a dual-authored poem in 1598, after Marlowe's death. Unlike Marlowe's lovers, Chapman's reach a happy ending. Hymen, like Leander, illicitly penetrates a well-born virgin's hymen, but unlike Marlowe's heroine, Chapman's experiences no shame or sadness after being penetrated. She seems not even to be conscious that her "incorruptible garrison" has been breached (*Hero and Leander*, Sestiad V. 253). For his conception of Hymen's androgynous persona, Chapman draws on Cartari, who explains that the beautiful Hymen carries a bridal veil in one hand (a visual allusion to the hymen) and a phallic torch in the other (see Fig. 2.1).

Cartari's (and his illustrator Bolognino Zaltieri's) image captures both a pre-nuptial moment, when the lovely and garlanded young man is carefree in a homosocial space, and a post-nuptial time when Himeneo passes young boys absorbed in playing with nuts that symbolize the testicles, a part of the man's anatomy that is "doubly protected," like a walnut with its interior nut. Walnuts, as Pliny the Elder explains in his *Natural History* (77 CE), to which Cartari alludes, have become "emblems consecrated to weddings."[16] The engraved image looks ahead to a post-nuptial moment when boy children have been born and are already playing with their own generative parts (*noci*) to continue their father's bloodline and the transfer of property across the generations. The image thus constitutes a set of visual-anatomical puns alluding both to male and female generative parts and, through the image of boys playing with nuts, to the male progeny that are the goal of both pagan and Christian marriage.[17] Cartari's image offers a visual-anatomical allegory pointing to Hymen's past, present, and future as a god who presides over the erotic joining of male and female genital parts.

What Cartari and Zaltieri's image of Himeneo cannot convey are the emotions, and sensations, of the bride and groom during the wedding. George Chapman was very interested in such emotions and sensations. He explores them in a tale told to distract Hero from her tragic wait for Leander. In Sestiad V,

Figure 2.1 Vincenzo Cartari, *Le imagini de i dei de gli antichi* (1571), showing Himeneo "walking beside some boys playing with nuts." Design of illustration attributed to Giuseppe Salviati (*c.* 1520–1575 Venice); etched by Bolognino Zaltieri (active Venice, 1555–76); published by Vincenzo Valgrisi, Venice. Image copyright © The Metropolitan Museum of Art. Image source: Art Resource, NY.

she is figured as enduring (and sponsoring) the wedding of one of her own servants—a ward like many of Elizabeth I's ladies in waiting—during the time when a nymph named Teras tells Hymen's tale. At this moment in the narrative, Hero has experienced only a "clandestine" marriage, with a loss of virginity she deeply regrets (see the end of Marlowe's Sestiad II). Chapman's Hymen aims to give his beloved a chance to experience the pleasures of a first erotic union.[18] The digression on Hymen provides a politically complex comic contrast to the tragic Leander; Chapman portrays Hymen's beloved Eucharis as enjoying the moment when her hymeneal threshold is penetrated by Hymen in a strange triune form: he merges with Cupid (Eros) and Proteus, the god of metamorphosis, to join with his beloved—whose name signifies "good grace" and perhaps alludes to the ceremony of the Eucharist that divided Protestants from Catholics during the Reformation. In Cartari's and Chapman's texts, and in many of Hymen's other early-modern guises, the god's genealogy, gender, sexual desires, and status as mortal or immortal being remain uncertain.[19]

By the first decade of the Reformation, the three main definitions of "hymen" circulating since the early Christian era were visible in print for vernacular readers to ponder, although we cannot know who saw them when or how they were interpreted together or separately. Sir Thomas Elyot, an intellectual servant of King Henry VIII, gives two definitions (one below the other) on a single page of his *Dictionary* (1538). The first is a medical-ideological statement defining "hymen" as "a skin in the secret place of a maiden, which, when she is deflowered, is broken."[20] Second is the terse statement that "Hymenaeus was said to be a god of marriage." The formulation, which distances the writer from the opinion, recalls Servius's statement about a so-called virginal membrane. Many pages later, tucked away in an appendix of Latin phrases and not noticed by the small band of "hymeneutics" scholars who have cited Elyot's *Dictionary*, is "*Concelebrare platea~hymenaeo:* to make joy abroad with singing and dancing in the honor of marriage."[21]

It is an irony of history that Elyot offered this third definition of a hymeneal marriage as a public celebration only a few years after he had contributed to the Latin documents that Henry VIII and his advisors sent to the Pope arguing for the English king's right to dissolve his marriage to Catherine of Aragon.[22] After twenty-five years of marriage and the birth of one living child—a girl—Henry no longer accepted his wife's claim that she had been "virgo intacta" (an intact virgin) on their wedding night. The Pope had accepted her belief and had given the couple a dispensation to marry; he refused to change his mind. No physical evidence remained, of course, to support Catherine's view not only of her own younger body but also of her first husband Arthur, who had been unable, she consistently said, to break her maidenhead during their marriage. The Church of England headed by a king—and later by the Roman Catholic Queen Mary (who couldn't get pregnant with her foreign husband Philip) and still later by another queen, Elizabeth (who declined to marry and subject her royal body to the dangers of childbirth and to the authority of a husband like the one who had killed her mother)—was a new institution threatened not only by ongoing religious disputes but also by a crisis of "bio-social reproduction," as Jennifer Summit puts it.[23] The crisis existed chiefly at the level of the monarchy, but it engaged the imaginations of many English subjects from the era of Henry's Acts of Succession (beginning in 1533) to the deposition of the last Stuart and Catholic-leaning monarch, James II, in 1689. Anatomical hymens were contested and sometimes manually tested evidence in legal cases of aristocratic divorce, as in the famous instance of Frances Howard's marriage in 1606 (aged 13) and subsequent divorce seven years later from the equally young Robert Devereux, third Earl of Essex. Repeated celebrations of Hymen as a pagan god conjured images of anatomical hymens, along with the problems of male impotence and the visions of erotic coupling in a "secret" place in the female body that these celebrations invoked.

Many studies of early modern religious doubt focus on the miraculous certainty granted to Christ's doubting disciple Thomas in relation to the wound on Christi's resurrected body (John 20).[24] Another small body of work has focused on Salome, the midwife, whose doubt about Mary's post-partum intactness was met by the withering of Salome's curious finger, inserted into Mary's vagina. Mary miraculously healed her, however, and according to the Gospel of James (150 CE, later deemed apocryphal), Salome returned to full belief, becoming a vivid figure in English mystery plays and in the liturgy for Christmas. Her testing of a doctrine of "perpetual virginity," which Calvin rejected, was largely suppressed after the Reformation.[25] In both Salome's and Thomas's cases, doubt is resolved through miracle by an unseen and divine process. Other studies of doubt in early modern science have focused on the technologies and social histories of knowledge acquisition, progressing, albeit sometimes slowly and against obstacles, to follow a complex inductive process.[26] Historians of science have often traced the rise of a theory of scientific knowledge-acquisition, captured by Francis Bacon in *The Advancement of Learning* (1605): "If a man will begin with certainties, he shall end with doubts; but if he will be content to begin with doubts, he shall end in certainties" (1.5.8).[27] The content of the religious and scientific "certainties" is of course different, but they share an ideal of linear progress leading from doubt to certainty.[28]

For Bacon and his followers, the inductive method led from doubt to certainty via replicable and mostly visible procedures. This essay, by contrast, brings together questions about religious miracles and scientific methodologies to cast new light on an underappreciated site in both literatures: the female body, the mode of being of "pagan" gods in Protestant territories, and the "threshold" to the womb. That threshold was beginning to be observed, like territories of the New World, by scientists intent on seeing what had previously been hidden in cultures that prohibited dissection of human corpses. This threshold was also associated with a god representing (albeit with notable exceptions) the procreative goal of marriage and the father's desire for certainty through a pure bloodline. But uncertainties about the anatomical hymen converged with a more general problem of cultural belief having to do with the pagan gods in the early modern era in Europe and England. Investigating this problem in *The Pagan Dream of the Renaissance*, Joscelyn Godwin writes that there was a "long period of European history in which no one believed in the existence of Jupiter, Juno, and their Olympic court" or in the existence of their Greek almost-doubles, Zeus and Hera. "'Believed,' however,"—Godwin adds—"and 'existence,' are loaded terms which do not always exclude their opposites."[29] I propose that the pagan god Hymen and the site of anatomical debate that bears this name illuminate, and are illuminated by, the questions about belief in the existence of unseen entities that arise from multiple translations of St. Paul's famous definition of faith in his letter to the Hebrews.

Belief, Doubt, and Questions of Evidence

In their translation of St. Paul's Letter to the Hebrews, a group of Protestant exiles in Geneva in the 1550s decided on a novel rendering of a word in Chapter 11, verse 1: the famous definition of faith quoted in my third epigraph above. They translated the verse as "Faith is the ground of things, which are hoped for, and the evidence of things, which are not seen." Their departure from what had been a multi-lingual understanding of the Greek word *elenchos*—which they render as "evidence"—might not be immediately striking to modern readers of the King James Version (1611), which follows the Genevan innovation. But in earlier versions of the English Bible such as John Wycliffe's (1380) and the influential one by William Tyndale (martyred by Henry VIII in 1535), the word for *elenchos* was "argument," closely following the "argumentum" given by St. Jerome in the Vulgate (fourth century CE). Calvin's cousin Olivétan, who translated the Bible into French, rendered *elenchus* as "certification" in 1535.[30] And later English translators suggest "conviction" rather than "evidence." Among them is the famous nineteenth-century *Concordance* compiler James Strong. Neither the Greek nor Latin equivalents for "evidence," nor Luther's quite wonderful *Nichtzweifeln* ("not doubting") of 1522, bring forward a semantic connection with acts of seeing. On the contrary, earlier and some later translators stress an affective state of *security*, not the doubt-inducing phenomenon of evidence as "that which comes from seeing."

For the Geneva Protestants, there is arguably a valuable challenge to be apprehended and overcome in pondering the implications of "evidence" in relation to an apophatic concept of faith in God as an invisible and finally unknowable Being. The resonances of this translation for historians of science, and, in particular, for what Aaron Kitch calls the "visual theologies" of English Protestant culture, have not, so far as I know, been discussed.[31] But the translation's injunction to give total credit to the paradoxical *evidence* of that which cannot be seen takes us directly into the realm of the miraculous explored by Lorraine Daston, who reminds us that "only when enlisted in the service of a claim or a conjecture do [facts] […] become evidence, or facts with significance." She continues: "evidence might be described as facts hammered into signposts, which point beyond themselves and their sheer, brute thingness to states of affairs to which we have no direct access."[32] The hymen has received little modern scientific attention at the cellular level as a mucous membrane. The alleged fact of a hymeneal membrane signifying virginity has been doubted over the centuries but also repeatedly "hammered into a signpost" that is particularly shaky because it is so often composed of words. Hymens are not depicted in early modern anatomical illustrations, although there is one illustration of the uterus in Andreas Vesalius's *De humani corporis fabrica libri septem* (On the Fabric of the Human Body in Seven Books, 1543) that Helkiah Crooke reproduces in *Mikrokosmographia* (1615), labeling the "hymen" as a tiny letter "o" in the "privitie." But there is no visible hymeneal part.[33]

The problems of generalizing from an evidentiary field consisting mainly of early modern physicians' and midwives' descriptions (the latter often scornfully reported by men) can be dramatized by juxtaposing two statements about evidence.[34] One is by a nineteenth-century physicist, Victor Stenger, a lapsed Roman Catholic who supported the thesis of an inevitable conflict between science and religion. "The totality of evidence," he wrote, "indicates that, on the whole, over the millennia the Christian religion was more of a hindrance than a help to the development of science."[35] The second statement is by Richard Baxter, a dissenting English Protestant whose radically liberal views on the doctrine of "atonement" differed from Calvin's in their stress on human agency.[36] In the preface to a treatise of 1682 on debates about the "divine authority" of the Mosaic books of the Bible, Baxter wrote:

> It is become [so] great [a] controversy, whether Faith have any *Evidence* or not, and whether we can certainly *prove* the Gospel to be true, or rather must merit the more by believing it without proof, that the Papists are together by the Ears about it; and those Protestants that handle it, differ among themselves. But the most keep their peace by not daring to decide it.[37]

Baxter arguably deepens the controversy by insisting that we don't know God's "original" words except through highly fallible human hands, and yet he seemingly accepts the Lutheran idea that our *only* evidence for faith in God is the Scripture—or rather, the written, then printed, scripture in all its many languages and textual variations. Baxter adheres, that is, to the Protestant doctrine of *sola scriptura*, but at the same time undermines it by remarking that the visible letters of scripture are not simply evidence to be believed. Instead, they are *difficult* evidence, what he calls *intelligible* evidence, which I take to mean requiring interpretation, because they themselves are translations and interpretations.

The conclusion of Baxter's argument takes "evidence" out of the realm of what is apprehensible through the senses but nonetheless insists that it can be known as *something other* than its invisible referent or signified: "how can those Teachers show the people the ascertaining Evidence and Proof," Baxter asks, "who hold that there is none to be shown. The objects of Faith are not evident to Sense, (not seen, not tasted, *etc.*) but the truth of the Revelation hath ascertaining proof: And nothing is provable but by intelligible Evidence."[38] Questions about what counted as credible or "intelligible" evidence for Christian writers and readers were clearly circulating among sixteenth-century anatomists concerned with the secret parts of the female body. An exquisitely poised statement about the hymen's existence as a sign of virginity comes from Vesalius, whose *Fabrica*, with its amazing illustrations, was influential throughout Europe. In an Appendix added in 1555, Vesalius writes: "If it is true that the hymen is broken and damaged at first coitus, I believe it provides no small or rare proof of praiseworthy virginity to some women before marriage."[39]

If, as Harry Newman suggests, a hymen construed as a virginal "seal" is like a wax security lock on a letter, then Vesalius's elegantly hypothetical statement offers a form of security to his readers; the hypothetical statement itself, as Touchstone observes in *As You Like It*, is a "peacemaker: much virtue in if" (5.4.94).[40] But the line between peacemaker and intellectual appeaser is thin. I want to consider two examples—one in Latin, by Vesalius, one in English, by Crooke—of an early modern medical attitude toward hymens that I call "dubious belief." This attitude, which has both affective and intellectual dimensions, arises in part from the author's anxieties about the proper limits of professional decorum on topics of human sexuality and reproduction. What does a medically educated investigator, writing for a mostly male readership, perceive or *envision* when he publishes words about having "found" a hymen on a dead or living female's body? Should words about such a thing (or "nothing") be made public at all, either in Latin, for an educated elite, or in a vernacular language? And what is the doctor's probing or writing hand to do in relation to theories held by others, among them midwives who had hands-on experience that in many cases the male physician did not?

Concerns about anatomical hymens pertain to evidence that necessarily requires confirmation over time with different bodies. The confirmation process requires acts of touching skin and bones directly or with prostheses that may enhance vision such as ancient mirror-probes.[41] Scientific seeing in medicine and dissection blends into some form of sensory contact which in turn blends uneasily, I surmise, into fears about opening and potentially *wounding* a female body, whether dead or alive, in a scene of knowledge-acquisition that conjures up specters of incest, rape, and illicit rivalry between the anatomist and other male figures. Vesalius depicts several such scenes of knowledge-acquisition in his *Epistola rationem modumque propinandi radicis Chynae decocti* (China Root Epistle, 1546). It is a strange and hybrid text addressed specifically to his physician friend Joachim Roelants. The title page advertises Vesalius's newly acquired position as physician to the Emperor Charles V, who had suffered "recently" from gout. So had another major patron, Duke Cosimo de' Medici, to whom there is a dedicatory letter. The epistle takes an herbal remedy, a boiled China root (*smilax China*) that was alleged to be effective not only for gout but also for syphilis and kidney stones, as a cover for defending the methods and findings of the *Fabrica* against a harsh attack by one of Vesalius's former teachers, Jacobus Sylvius. He despised Vesalius's departures from Galenic truth in the *Fabrica* and indeed proclaimed Galen's authority to be infallible, as if he were the Pope of medicine.[42] Vesalius's title page promises a "summary" of a letter to Sylvius that will be "of Great Usefulness to Students of the Truth [...] Since it Shows How Easily Galen Has Heretofore Been Excessively Trusted on that Subject" (i.e., the truth of anatomy).[43] Vesalius's counterattack on Sylvius creates a rhetorical arena fraught with Oedipal tensions and professional anxieties about offending powerful men whose need for medical help reveals a chink in the armor of their patriarchal masculinity. It is hard for the modern reader to interpret Vesalius's tone, as he addresses readers variously presumed to be friendly, known to be hostile, and, in two notable cases, high above the author in the social hierarchy. His prose is convoluted: it requires but also rewards slow reading.

The text's repeated attacks on Galen's reliance on the dissection of animal bodies show not only Vesalius's pride in the methods and conclusions of his *Fabrica* but also his interpretation of the history of classical medicine. Which of the ancient medical discourses best represents the "whole" to be admired by early modern humanists (for that is how Sylvius defined himself and his pupils)? The historiographical question, with its focus on parts representing wholes, has a particular resonance for the developing science of anatomy. In the eyes of Sylvius, Galen (second century CE) represented the cumulative wisdom of ancient Greek medicine from the Hippocratic corpus onward. For Vesalius, by contrast, the Alexandrian anatomists of the third century BCE, Herophilus and Erasistratus, were the significant representatives of classical medical wisdom because they drew conclusions (many of them now lost) from dissections of human bodies. For modern historians of science such as Heinrich Von Staden, however, the Alexandrian anatomists constitute a brief historical exception to the general ancient Greek rule, which involved a long-standing taboo on human dissection.[44] Competing ideas about innovation in medical practices, about Arabic textual authority in contrast to Greek, and about the humanist concept of *mimesis*, which informed ideas about which authorities are worthy of *imitation,* underlie Vesalius's *Epistola*.[45] Its narrative about his quest for certainty about hymeneal membranes illustrates Thomas Kuhn's view that science does not progress in a linear fashion toward greater truths.[46]

"For the Sake of the Hymen"

Vesalius begins his narration of his quest for knowledge about the brave new world inside the female body by explaining why he was silent about hymens in his *Fabrica*. He was "purposefully silent about the hymen," he writes, using a Latin phrase (*ex industria*) and drawing an implicit contrast between his self-disciplined silence and Galen's, which derived from ignorance.[47] He adds another reason for his silence: he then "knew nothing with certainty," having "never dissected a virgin" (G 166, L 139). Suddenly, however, he admits that he had in truth dissected a virgin's body—that of a six-year-old girl whose corpse was unlawfully procured—before publishing the *Fabrica*: "I had gotten [the child's body] for preparation of a skeleton with the help of a student at Padua who had stolen her from a grave" (G 166, L 139). While he was

> cutting everything away from her bones [...] and not taking the time to inspect any part, I did however dissect the uterus close to the hymen. Though I found it as I have recently seen it since the publication of my books, I did not dare to say anything about it because I perceived that animals do not have a hymen. So I never take a position on the basis of one dissection or another.
>
> (G 167, L 139)

This admirable scientific principle about not generalizing on the basis of just one experiment precedes Vesalius's admission that he had initially perceived the child's hymen as belonging to an anatomical category of "connective tissue"—Aristotle's and Galen's understanding of the substance. Vesalius acknowledges to medically trained readers that he had not really "seen" the hymen as a mark of female virginity at all. He also mentions (with a rhetorical gesture of "passing over," *praeteritio*) a corpus of Arabic medical authorities who understand the hymeneal tissue as a *porous network* of veins, a *cento* or "patchwork" (G 167, L 139) rather than a singular seal across the mouth of the vagina.

Vesalius relates the stories of four more dissections of bodies he thought belonged to virgins: two of these, a hunchback girl whom he thought no man would have desired and a thirty-six-year-old nun, were procured for him by Cosimo de' Medici. A third belonged to an "elegant" prostitute. The fourth and most epistemologically complex case was that of an eighteen-year-old noblewoman whose uncle desired an autopsy because he suspected that she had been poisoned. Vesalius undertook the work, he writes, "for the sake of the hymen" (*hymenaei occasione*).[48]

He performed a dissection of her uterus with one other doctor after most of the spectators had left the room. Like a good detective, he hints that he had suspected female "secrets" at work in the cause of death, which was likely neither from poison nor from the too-tight corset that he initially thought had contributed to the girl's respiratory difficulties at the end of her life. With a comic touch that shows the reader how weighty his opinions were—the "attendant women had left to shed their corsets as quickly as possible"—Vesalius writes ambiguously about finding a hymen, but one that was either "not whole" or "not obvious" to him.

His description of the noblewoman's body invites Giulia Sissa's question about the ontology as well as the epistemology of the anatomical hymen. Is it, Sissa asks, "perhaps a hypostasis, a fetish, an article of faith"?[49] Of his act of cutting into the noblewoman's body, Vesalius writes: "*At non fuit mihi hymeneus integre conspicuus*" (L 142). D. H. Garrison interprets "*non* [...] *integre*" as signifying "not obvious to me" (169), whereas C. D. O'Malley assumes that Vesalius perceived the hymen as "not entirely whole" (63). O'Malley thus emphasizes the condition of the putative *object* rather than that of the perceiving subject. Both are plausible readings, and in either case, the sight of (something like) a broken hymen did not provide satisfying evidence of the diagnosis Vesalius suspected: "strangulation of the uterus." This was a condition often discussed by Hippocratic and Galenic physicians and thought to involve a womb ("matrix") "wandering" upward in the woman's body.[50] In fact, Vesalius did find some evidence to support his theory: the girl's "hymen [...] was not entirely whole but had not entirely disappeared, as I have found is usually the case in female cadavers in which one can barely find the place where it had been. It looked as if the girl had ripped the hymen with her fingers either for some frivolous [*petulanter*] reason or according to a remedy from Rhazes" that cured strangulation of the uterus "without intervention from a man."[51] Rhazes (Abū Bakr Muhammad ibn Zakariyyā al-Rāzī) was a ninth-century CE Arab-Persian physician who suggested that womb wandering could be cured by a midwife (*obstetrix*) putting her oiled finger inside the vagina to massage it.[52] Such a massage, which Rhazes recommends for times when the superior remedy of intercourse with a husband is not available, would have likely broken a hymen construed in the Christian tropes of a seal or a locked gate—as would the girl's own finger-actions, which Vesalius, mediated by O'Malley, envisions as "ripping" the vagina for some "frivolous reason."[53] Marie Loughlin suggests that Vesalius is alluding here to self-pleasuring masturbation.[54] On the occasion of this dissection, Vesalius didn't find enough to help him determine a clear cause for the girl's death or to reveal what motivated his quest for precise material knowledge about her hymen's mode of existence after any hope for a cure had passed. Is he writing about a *partly* broken hymen (the entity imagined as an "integral" seal) that undergoes a natural process of "disappearing" over time or about a membrane that was not fully visible when examined because it was "hidden," as it often is in young girls (however we define "often")?[55] The passage gives an inconclusive sketch of a dead girl's inner body and probes the thin ethical line between professional dissection and illicit mutilation of a body. Vesalius also considers, albeit disapprovingly, the possibility that a noble woman "ripped" her own inner parts in anger or in a quest for pleasure. *Petulanter* can mean not only "frivolous" but "wanton."

Vesalius's illustration of doubting belief in a hymen's existence comes, as I've suggested, both from his elegantly hypothetical sentence in the Appendix of the *Fabrica's* second edition (1555) and from the enigmatic sequence of descriptions about his "findings" of hymeneal membranes in the earlier *Epistola*. The most thoughtful English commentary on hymens occurs in Crooke's *Mikrokosmographia*. He titles Chapter 15 of Book 4 "Of the neck of the womb & the hymen"; and he returns to the hymen and the trope of the womb's "neck" in "Question 13" in the section called "Controversies" that concludes this and the other books of his long compendium of medical opinions. Like other early modern English anatomists, he uses many illustrations from Vesalius's *Fabrica*, and it is often hard to distinguish Crooke's opinions from those of earlier English anatomists and others writing in many languages. The twists and turns of Crooke's prose, peppered with foreign names and terms, are arguably colored by his desire to avoid censorship. Book 4, which focuses on human generation, had already been threatened by the Bishop of London, John King, on the basis of his reading

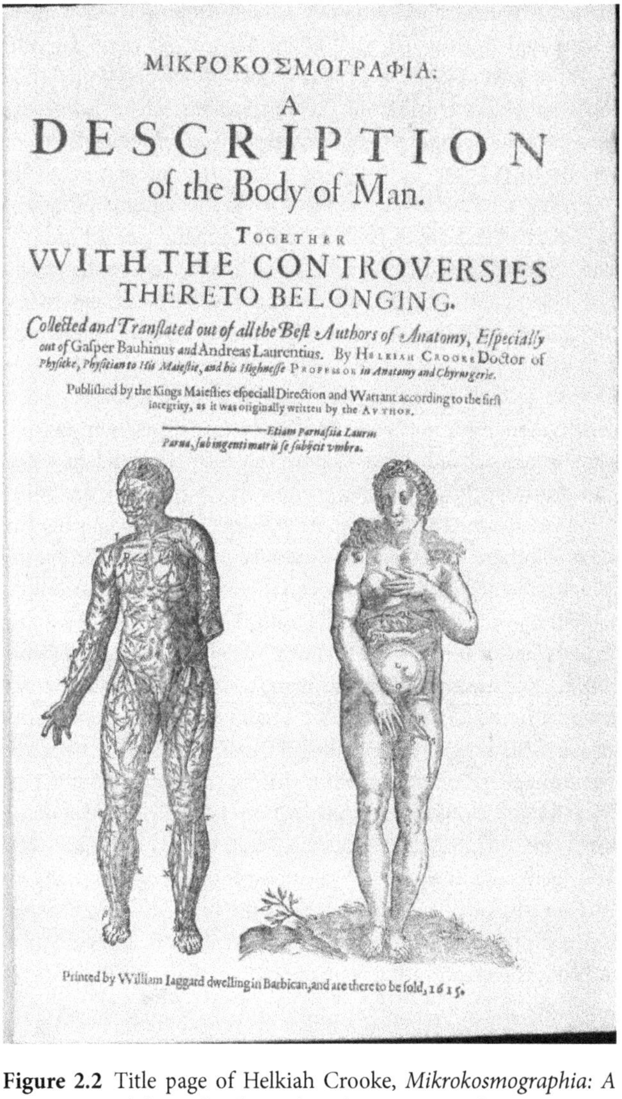

Figure 2.2 Title page of Helkiah Crooke, *Mikrokosmographia: A Description of the Body of Man* (1615). RB 53985. The Huntington Library, San Marino, California.

of a pre-publication copy. Crooke insists on his 1615 title page that his book was published in its "integrity" (see Fig. 2.2).

Crooke offers an eloquent defense against censorship (or the kind of self-silencing that Vesalius had described in his *Epistola*) in the preface to Book 4, the one that had most offended the Bishop of London.[56] Sharing the common view of the genitals as shameful, the Bishop tried to get the College of Physicians and the publisher, William Jaggard, to omit the offending chapters. At the start of Book 4, Crooke reports a conversation with himself in which he considers cutting out the ensuing part of his book—but that would have been like amputating a limb from God's creation.[57] The human body in all its parts is the "epitome" of divine creation, and man's "seed"—among the other topics taken up in Book 4—is the epitome of the epitome. The valued part-object shifts from being an appendage, a limb, to being an essence, something critical to life. The self-reflective sentences give the prospect of cutting parts of both bodies and books a psychoanalytic charge that spills into the discussion of the hymen as a "mark of virginity" in Chapter 15 and in the book's final chapter ("Controversy" 13).

For Crooke, as for Vesalius, the physician's quest for knowledge about the existence of hymens leads not to certainty but instead to proliferating questions about what is natural and what is not. The former would be God-given and would pre-exist the physician's investigation; the latter, an "unnatural" hymen, could be deemed a fiction, as we will see Crooke suggesting (through the words of another); or it could be deemed a pathological condition rather than a body part with which every woman is born. Another possibility, which blurs the distinction between natural and unnatural, is that the anatomical hymen is an allegorical interpretation or a socially inflected construction—in this respect like the ancient god Hymen. Vesalius and Crooke sometimes seem to believe in the idea of the hymen as a human-made construction, and sometimes, like Shakespeare's Orlando, they seem to oscillate between states of belief and unbelief. For both medical writers, the hymeneal entity emerges in rhetorical formulations that blur distinctions not only between natural and unnatural bodily phenomena but also between objects of inquiry gendered female and agents of knowledge-acquisition gendered male, scientific agents who sometimes identify with the figure of the husband whose penetration of the vagina destroys what might exist, or have existed, as a somatic sign of virginity. What is seen and what is imagined?

Natural, Unnatural, Seen, or Unseen

Crooke mentions a number of European medical authorities who consider the "natural" hymen a fiction, but he doesn't commit to this opinion himself. He also mentions the concept of an "unnatural" (yet "organical") "imperforate" hymen. He gives the word "*atresia*," literally, without perforation, in Greek letters as well as in the Latin "*imperforata*"; this occurs in his description of women who have the problem of an obstructed vagina (Book 4, Controversy 13, 256). Perhaps flaunting his classical learning,

Crooke mentions Soranus of Ephesus (235), but not in the chapter about "imperforate" hymens. Soranus, as we saw earlier, uses the term "hymen" for a specifically female body part, as Aristotle and Galen did not; but Soranus does so only to define it as a superstitious fiction. He denies the material existence of a hymeneal sign of virginity in the same century that the midwife Salome was represented as being miraculously both punished and healed after doubting the newly formulated doctrine of Mary's perpetual virginity.[58] For Soranus, the "hymen" is not a thing but a false belief about the female body. Soranus specifically advises his ideal readers, midwives literate in Greek, to be "free from superstition" and skeptical of "vulgar rites."[59] The manual examination of Christian virgins may have been among them. Early church fathers of the fourth century CE such as Ambrose and Augustine decry those officials and midwives who search for the somatic evidence of virginity while neglecting the woman's *invisible* spiritual quality of sanctity. In his *City of God*, Augustine writes that the body's holiness (*corpus sanctum*) "lies not in the integrity of its parts nor in the fact that they are not violated by contact, for in various accidents they are liable to suffer violence [...]. A midwife, as she investigated the maidenhead of a certain virgin, destroyed it [with her hand] either by ill intent or clumsiness or accident during examination" (*Obstetrix virginis cuiusdam integritatem manu velut explorans sive malevolentia sive inscitia sive casu, dum inspicit, perdidit*).[60]

Soranus clearly distinguishes a false belief in a virginal membrane (normal or miraculous) from the true problem of "*atresia*." This rare condition, he maintains, is often erroneously confused with the fictional hymen (*Gynecology*, trans. Temkin, 15). It requires medical removal in order for the woman to receive seed and bear a child. Soranus was not a Christian, so far as we know, but he is one of the few physicians on record ever to regard virginity as a more "healthful" state than marriage is.[61] Crooke names Soranus in a section from Book 4, chapter 15 entitled "Of the Neck of the Womb and of the Hymen" that at once echoes some of Soranus's views and contradicts his major argument about the non-existence of the "hymen" believed to grow "across the vagina, dividing it," and causing pain "when it bursts in defloration." Crooke repeats Soranus's view that what "bursts" in defloration is not a seal or veil-like thing covering the vaginal opening but instead some small blood vessels and "folds":

> The attrition of these folds and their extension in the first society of maids with men, *Soranus* thought to be the cause of some maidens' pain in devirgination or loss of their maiden-head as we speak, and because certain veins pass by them these being broken by the husband the blood issueth sometimes in great abundance.
>
> (235)

This statement, as Kathleen Coyne Kelly suggests, defines the hymen as a *site,* not as a discrete *structure.*[62] But Soranus's view of folds and vessels is partially, although not fully, contradicted when Crooke reassures the English Protestant reader about the existence of a seal that a husband could perceive himself as breaking as he enters the "neck" of the vagina. "Many," Crooke writes, "will have [the hymen] to be a slender membrane," placed "overthwart that it may shut the cavity of the neck of the womb" (235). "May" acquires the force of "perhaps" as the exposition continues, however, undermining the idea of a singular seal and also the idea that a penis rather than a finger would "know" the hymen as it takes on the form of a membrane already perforated (just a little bit) before a nuptial penetration occurs: "Yet," Crooke writes, the hymen is "perforated in the midst like a ring, that in grown maids it will admit the top of a little finger, that through it the [menstrual] courses may passe" (235). Anatomical nuance abounds, but so does deniability. It is deftly created by syntax, diction, and a cascade of tropes.

In the final chapter of Book 4, Crooke plays on the word "certain" as signifying both "some" and "sure": "It hath been an old question and so continueth to this day, whether there be any certain marks or notes of virginity in women and what they are" (255). Although he insists at first that there is ("almost") unanimity on the "certain" existence of a hymeneal membrane, in the same sentence, he demonstrates the lack of expert consensus on where it is "placed": some say it is "in the midst" of the "neck of the womb," others say it is right "after" (next to?) the opening for urination. Placing the reader in a perfect limbo, Crooke gestures toward an anonymous but biblically authorized group ("they say") who think that the hymen exists across the vagina and "is broken or torn in their [virgins'] first accompanying with men; and therefore they call it *Claustrum virginitatis, The lock of virginity:* for which their opinion they bring testimonies out of the holy scriptures."[63] He then cites three authorities who hold two different views: one, Fallopius, "yields" to the pseudo-biblical view, pseudo because the passage adduced from the Hebrew Scriptures (Deut. 22:13-21) has to do with post-nuptial blood as a "token" of a new wife's sexual purity before marriage, which a new husband may publicly doubt. The passage does not show any knowledge of a somatic hymen. A second authority (Columbus) is reported as writing that "he hath seen it [the hymen]." A third authority, who is given the most rhetorical room, not only says he hasn't seen it but conjures up a violent spectacle of repeated investigations of very young girls' dead bodies:

> *Laurentius* sayeth that he hath cut up maiden children born before their time, of three months, of 3, 4, 6, and 7 years old, and yet he could never find it though he searched curiously for it with a Probe; which (sayeth

> he) might have been felt to resist the Probe if there had been any such thing, and therefore he thinketh that it is but a mere fable.
>
> (255–6)

Two out of three authorities testify to the hymen's material existence (one by "yielding" to someone else's opinion); but the third authority, last but not least, prompts Crooke to disavow his own metaphor of the "lock" of virginity, which had fleetingly constructed the husband who breaks it as a thief, and which arguably applies both to the female body and to the epistemological situation as the physician has created it. He denies the validity of the "lock" metaphor just after he moves across a rather slippery border between Laurentius's denial of a "natural" hymen and his concession to others that there is another, "unnatural" membrane, rare, but still "organical." The rhetorical drama arises as if it were a debate about a map, among experts, one of whom "partially" believes in others' views of the new territory:

> Yet notwithstanding, [Laurentius] thus far giveth credit to *Columbus* and *Falopius,* that he thinketh there is sometimes such a membrane found; but if it be stretched overthwart in the middle or at the end of the neck of the womb, then he thinketh it is not Natural, but an Organical disease or [the result] of the instrument being faulty in conformation.
>
> (256)

The "instrument" is not the physician's probe but the human female's body, and if it is "faulty in conformation," it is so because it can't perform its natural function of conceiving and bearing a man's child. Granting the existence of a membrane that looks like a hymen but is not a "natural" barrier because it can't be broken by intercourse, and granting also that some physicians do not believe in the existence of this diseased membrane at all, Crooke concludes that neither atresia nor the idea of a natural hymen constitutes a satisfactory sign of female virginity. "We must therefore," he writes, "find out some other lock of Virginity" (256). Crooke thus follows Soranus on the existence of atresia but does not repeat Soranus's view that instances of atresia—which are real and need medical attention—should *not* be understood as "thick" versions of a "natural" hymen. Crooke omits, and perhaps censors, Soranus's claim that a "natural" hymen is a fiction and Crooke also omits Soranus's point that "imperforate" (that is, unnatural) vaginal membranes ("hymens" in the Aristotelian sense of the word) should not be seen as invariant body "parts." These, for Soranus, always occur in the same place across the bodies the physician examines: a hand is always at the end of an arm, not a leg.[64] The same objection could be made against the idea of the "natural" hymen, which both midwives and physicians located in different parts of the womb's "neck."

Although Crooke (like many modern medical experts) does not follow Soranus in drawing a clear distinction between an imperforate hymen and the (idea of) a "natural" one, his account of a vaginal obstruction that an expert must remove helps explain the "membrana" that Ben Jonson attributed to Elizabeth I, according to Kaara Peterson's reading of a famous crux in a letter that Jonson wrote tendentiously after Elizabeth's death. Supposedly drawing on information from the late queen's own physicians, Jonson wrote in a letter to his friend William Drummond that she "had a membrana on her which made her uncapable of man, though for her delight she tried many."[65] In the cultural imaginary of early modern Protestantism, the concept of "imperforate hymen" belongs to a dyad that evidently requires physicians or husbands to "perforate" it as soon as possible, even though physicians cast doubt on its existence.

For devout Roman Catholics, this ideological dyad fails to allow for a valued and enduring state of virginity possessed both by ordinary Christian nuns and, supernaturally, by Christ's mother.[66] The dyad, which Crooke partly deconstructs by imagining the little "hole" in the hymen for blood to escape from the "enclosed" reproductive body, makes it easy to imagine an "unnatural" hymen as simply an excessively thick or hard version of a natural one. Soranus attempts to correct midwives who believe there is one hymeneal membrane, thin or thick, and that it bursts bloodily upon defloration. Soranus's near-contemporary Galen never mentions such a membrane. Writing when the doctrine of the "perpetual" virginity of Christ's mother Mary had not yet been fully articulated, although her miraculous virginity before Christ's birth was later granted by Calvin and was a source of wonder, Augustine in the fourth century had praised Soranus as a great physician. He may not, however, have read the gynecological treatise: he does not accept its major argument about the somatic hymen being a "*pseudos*." I suspect Crooke had read some Greek version of Soranus's text, enough to quote the word "*atresia*," which other early modern anatomists do not mention at all (so far as I know). But evidence of such reading is lacking, and there remain many gaps in the reception history of Soranus's text. Whether through prudence, or through disagreement, or through an imperfect understanding of Soranus's points about "pseudo" hymens and about atresia, Crooke offers his readers multiple versions of hymens and withholds full credit from them all.

Conclusion

This essay has explored a species of doubt that sometimes arises and sometimes does not in what we might regard as "cool" zones of Christian religious belief: among them are the London theater, the spaces of licit or illicit anatomical dissection described by Vesalius, the book publishers like William Jaggard who printed Crooke's big book despite the Bishop of London's objections, and who later published the First Folio by Shakespeare. Anatomical hymens, representations of Hymen the god by Vincenzo Cartari, his illustrator, and the seldom-read George Chapman, along with a much better known "embodiment" of Hymen, as Susanne L. Wofford calls it, in a play text that was "stayed" (censored) by the Stationer's Register in 1600—these historical phenomena emerge for modern scholars' inspection in a zone of doubt that continues to exist today, although one likely below the radar of many. It does so, for example, in the form of what Katherine Eggert calls "disknowledge" or willed ignorance.[67] It also does so in the practice of surgical operations to "replace" hymeneal membranes that are believed to be "signs" of virginity by men who have the power to kill women considered unchaste—and by men and women who desire to experience a "refreshment" of their marriage through a reenactment of a wedding night defloration scene. Hymenoplasties are the only kind of plastic surgery I know of that have a short shelf life because an inserted pouch of blood will not last long. Re-virgination efforts are not new; there are medieval recipes for accomplishing the appearance of virginity, and there are a number of Renaissance dramas—Thomas Middleton's *The Changeling* (1622), for example—that represent ways that a woman can pass a virginity test without having the invisible quality of sexual purity. But the global incidence today of hymenoplasty surgery supports Giulia Sissa's contention that "Hymen is a problem, still," for medical science and for the health of the body politic.[68] Sissa's question bears repeating: is the somatic hymen "a hypostasis, a fetish, an article of faith"?[69] In its continuing existence as an item of (some peoples') consequential belief, the anatomical hymen inhabits a border territory between religion and science that is mediated, and sometimes interrogated, by texts now classified as literary. Cultural and literary historians could well pay more attention to stories about hymens told today by forensic scientists, writers of health textbooks for teenagers, and TV shows such as *True Blood*, which dramatizes a heroine whose hymen regenerates each time it is painfully broken. If the anatomical hymen is a social construction that inspires doubt about individual women (is *she* pure?) and about human female bodies in general, perhaps it is time to ask, as many have before me, who profits (*cui bono*) from certain long-enduring beliefs, ones that seem intertwined with but also strangely invulnerable to doubt? In the light of Peter Harrison's contrast, quoted below, between scientific realism and "instrumentalism," I ask, in closing, for whom do hymens continue to be useful? "While many scientists," Harrison writes, "are themselves realists, in the sense that they believe that their theories and models depict reality in increasingly precise ways, there are powerful arguments for instrumentalism, according to which our scientific models are better viewed as fictional devices for making predictions about the behaviors of unseen realities. The relevant criterion for instrumentalism is not truth, but usefulness."[70] A criterion of usefulness is more clearly time-bound than is a criterion of truth, however, and it is perhaps time to ask whether the branch of modern medical science named "gynecology"—the logos or discourse of women—could relegate the idea of the anatomical hymen, so deeply embedded in Christian ideologies of gender and marriage, to a domain of historical hypotheses that invite speculation and critique, but not belief?

Notes

1 All quotations of Shakespeare are from *The Norton Shakespeare*, 3d ed., ed. Stephen Greenblatt et al, hereafter cited parenthetically by act, scene, and line number. My third epigraph is slightly modernized from the facsimile of the Geneva Bible published in 2007 with an Introduction by L. Berry. The Letter to the Hebrews is attributed to Paul's follower Timothy in the King James Bible. My essay is indebted to many scholars (named but not adequately thanked in my notes), and particularly to my editors Aaron Kitch and Jennifer Rust. Thanks also to David Simpson, Susanna Ferguson, Jean Howard, Helen Cushman, Zack Lesser, Tom Bishop, Karen Newman, and David Hillman for valuable suggestions.

2 On the significance of "Ganymede" as a "sweet" boy and the story of his rape by Jove, in the form of an eagle, see Jeffrey Masten, *Queer Philologies*, 75, 132, and *passim*. Drawing on Leonard Barkan's study of Ganymede in the European Renaissance (*Transuming Passion*), Masten emphasizes the "ubiquity" of the "sign" of Ganymede in Elizabethan culture, where a London book stall was named the "Eagle and the Child." On the social risks of depicting homoerotic love in Protestant England, see Mary Bly, *Queer Virgins and Virgin Queans*, 87–9. She discusses Richard Barnfield's defense of the poet's love for "Ganymede" in *The Affectionate Shepherd* (1594). In that same year, in *Edward II*, Marlowe represented Queen Isabella's bitter wish that the "cup of Hymen" on her wedding day had contained poison. If it had, she would not have experienced being abandoned by her husband. He, faithless, desired his male favorite Gaveston even more than Jove loved Ganymede, she says (1.4.273-74).

3 For an illuminating analysis of Shakespeare's debt to Renaissance Italian pastoral dramas in *As You Like It*'s concluding scene with Hymen, see Wofford, "Hymen and the Gods on Stage." Wofford argues that ancient gods of marriage are represented in Italian pastoral dramas in ways that help us see that Shakespeare's "Hymen can be read and performed as embodying an actual power or a force associated through the fictional metaphor with powers of the gods but distinctively theatrical not religious." But what does

"religious" (85) mean here, considered in opposition to a set of aesthetic and meta-aesthetic effects called "theatrical"? I see the boundary as blurred.

4 On "strange" and "foreign" words in Renaissance England, see Wofford, "Foreign," 478–92.

5 See the address to Hymen as lord of the universe and "binder" of its seeds that begins Martianus Capella's influential *Marriage of Mercury and Philology* (5th c. CE). See also Katie Reid's recent study of Martianus's influence, through print, on a wider range of Renaissance readers throughout Europe than scholars had previously supposed. On Hymen as a sign of male impotence—something rarely considered in the critical literature on his godhood—see book X of Ovid's *Metamorphoses*, which begins with Hymen arriving with a "sputtering" torch to the marriage of Orpheus and Eurydice (lines 6–7). He comes "in vain" (lines 2–3), unable to bring any good omen to the wedding or to prevent the bride's early death. See Ovid, *Metamorphoses*, ed. and trans. F. J. Miller, vol. 2, 64–5, lines 1–10.

6 See Bakhtin, *The Dialogic Imagination*, trans. Holquist and Emerson, 84, 98, 248, and passim. Bakhtin borrows the idea of "chronotope" "almost as a metaphor" from mathematics and Einstein's theory of relativity.

7 On January 25, 2022, members of the British Parliament introduced an Amendment to the current Health and Care Act that would ban hymenoplasties altogether, along with the practice of "virginity testing." The Amendment passed, but I do not know how or if such a ban is being enforced.

8 On the Swedish ban on the term "hymen," see Sissa, "The Hymen Is a Problem Still," 69–70.

9 The hymen is absent from Thomas Laqueur's *Making Sex* and Jonathan Sawday's *The Body Emblazoned*. It is also omitted from David Hillman's and Carlo Mazzio's edited collection, *The Body in Parts*.

10 See Sissa, *Greek Virginity*, 66–7, 110–11, on uses of "hymen" in Aristotle's zoological tracts.

11 On the use of the term *diakoreusai*—literally, unmaidening (from *koré*)—in Soranus of Ephesus's *Gynaikeia*, and on other ancient words for a female's first experience of penetration by a male member, see Sissa, *Greek Virginity*, 77–8. Unlike English and the romance languages, Greek does not insist, in this word at least, that the virginal woman's body is like a flower (e.g., Rosalind, "rosa linda," lovely rose). Hymen, as Wofford observes ("Hymen and the Gods on Stage," 73), is absent from Shakespeare's main English source, Thomas Lodge's *Rosalynde* (1590).

12 See Freud, "The Taboo of Virginity," 201.

13 Sissa, 108, quoting from Servius's commentary on Virgil's *Aeneid* (4.99)—a passage in which Juno and Venus briefly imagine an alternative, goddess-ruled universe in which Aeneas and Dido are peacefully married instead of tragically separated by the gods.

14 Cited and translated in Sissa, 107.

15 See Pindar's *Threnodies* 3, in *Fragments*, ed. and translated by Race, 374–7.

16 See Cartari, *Le imagini de i dei de gli antichi*, 198. His allusion to Pliny the Elder (*Natural History*, 15:24, cited from the Loeb edition, trans. bk. 15. 24:348-49), takes us to several theories about the significance of the nuts (*nuces iuglans*), including that they were part of the bawdy all-male joking (*Fescennine*) part of the wedding: the nuts in this context refer to the male lovers the groom is told he must relinquish as "unlawful" after marriage, as Catullus insists in his well-known wedding epithalamion, *Carmina* 61:114-34 (in lines censored in both English and Latin in the Loeb bilingual edition). For the full text with the Latin refrain "concubine, nuces da" and English translation, see https://en.wikisource.org/wiki/Translation:Catullus_61. Pliny also alludes to the theory that the nuts were dropped loudly outside the nuptial chamber and thus drowned out the bride's cries. I cite Cartari's essay on "Himeneo" from the 1571 edition in the facsimile published by Garland Press as part of a series edited by Stephen Orgel. Cartari gives an extensive table of contents and compares different interpretations of the gods' stories; his introduction notes that only the Jews, among the ancients, followed the "true religion" and eschewed images. He describes with interest the customs of earlier peoples in the Mediterranean basin, as if he were reporting on the peoples of the Americas or the Indies.

17 On the patrilinear goal (*telos*) of producing progeny, along with a "non-utilitarian" (but hardly non-ideological) ideal of paternal "happiness"—viewing a long line of descendants stretching into the future and fulfilling the biblical injunction to increase and multiply—as articulated in early Christian marriage theory, see Michel Foucault, *Confessions of the Flesh*, 15–16. See also Simon Goldhill, *Foucault's Virginity*, 121, on the belief in the truth of virginity as signified by an "intact" bodily hymen. Such belief, Goldhill writes, testifies to "patriarchal concern with the knowability" of female sexual purity.

18 The painless breaching of Eucharis's virginal "garrison" (V.253-55) prompts a choral singing of the traditional "io hymen" wedding song by Eucharis, Hymen, and "Love" (V.253-55). See Marlowe, *Collected Poems*, ed. Cheney and Striar, 252. Although the poem still appears under Marlowe's name in some modern editions, the first edition of 1598 states clearly that it was "begun" by Marlowe and "finished" by Chapman. His completion of Marlowe's minor epic shows the invented characters of Hymen and Eucharis joining their voices not to engender a child but instead to make a "plot" (V.265). This involves delaying their public marriage in Athens until the father of Eucharis agrees to accept the lower-class Hymen as a son-in-law who will become a deity.

19 See Wofford, "Hymen and the Gods on Stage," 81, for a reproduction and discussion of the recently restored painting by Nicholas Poussin (1634) showing "Hymen Disguised as a Woman During an Offering to Priapus."

Poussin's Hymen looks like a dancing maenad grasping Priapus's large penis. See also Marcia Pointon, *Strategies for Showing*, on Poussin's painting and other visual representations of Hymen (191).

20 On the connotations of "secret," see Park, *Secrets of Women*, 26 and *passim*. See also Ferguson, "Hymeneal Instruction," 97–130.

21 The *Dictionary* of Sir Thomas Elyot, sig. Kii and "The Additions" (n.p.). See Marie Loughlin for a valuable overview of the scholarship in the field she calls "hymeneutics" in her book's title. See also Kelly, *Performing Virginity and Testing Chastity in the Middle Ages*, for medieval and Renaissance references to a variable concept of a hymeneal membrane across medical, religious, legal, and literary writings. "Hymen" is rendered in Latin both as "pellicula" and as "integritas."

22 For an astute discussion of the texts related to the King's "great matter," among them a late play by Shakespeare and Fletcher on Henry VIII subtitled *All Is True*, see Frances E. Dolan, *True Relations*, chapter 6.

23 Summit, *Lost Property*, 99. The long-durational reproductive crisis she hypothesizes encompassed the "Warming Pan Scandal" of 1688, which blocked the infant born to James II and his Catholic wife Catherine of Modena from inheriting the throne.

24 See, for example, Glenn W. Most, *Doubting Thomas*, and Barbara Pitkin, "The Reformation of Doubt."

25 On the narrative of the Salome story in the *Protevangelium Jacobi* (*c.* 150 CE), and on its rich afterlife in English religious drama and in the Roman Catholic liturgy, see Helen Cushman, "Handling Knowledge"; see also Naomi Seidman's discussion of the *Protevangelium* passage, for which she provides a translation, in the context of early debates among Christians and Jews about Mary's virginity, in *Faithful Renderings, 67.*

26 See especially Chapter 2, "Seeing and Believing," in Simon Schaffer and Steven Shapin's *Leviathan and the Air-Pump*. See also Deborah Harkness, *The Jewel House*, especially "A Note About 'Science,'" xv–xviii, "Prelude," 7–10, and 234, on Hugh Plat's way of marking the presence of doubt and contingency in his accounts of his experimental results.

27 Bacon, *The Advancement of Learning*, ed. Wright, 41.

28 This formulation does not do justice to theorists like Thomas Kuhn, who as the editors of this volume remark in their Introduction, argues against a notion of linear progress in science; nor does my formulation do justice to instances when religious doubts lead not to belief but to atheism, or representations thereof. See, e.g., Tiffany Jo Werth, "Atheist, Adulterer, Sodomite, Thief, Murderer, Lyer, Perjurer, Witch, Conjuror, Brute Beast?"

29 Godwin, *The Pagan Dream of the Renaissance*, 1.

30 Olivétan Bible (1535), sig. mmv. A digital copy of the text is available at https://archive.org/details/1535-olivetan-bible/page/n773/mode/2up.

31 See Kitch, *Shakespeare's Theater of Nature*, 206–9; he also explores tensions between Protestant distrust of images and the hope that—correctly used

and interpreted—they could help in reforming error. "Sight transforms seeing into types of believing" (194).

32 Daston, "Marvelous Facts and Miraculous Evidence," 93.

33 See Crooke, *Mikrokosmographia*, 210. This and all subsequent references are to 1615 edition of this text, which is discussed in detail below. Loughlin also reproduces the illustration of the uterus in *Hymeneutics*, 210.

34 See Caroline Bicks, *Midwiving Subjects in Shakespeare's England*; and Eve Keller, "The Subject of Touch: Medical Authority in Early Modern Midwifery," in *Sensible Flesh*, ed. Harvey, 62–80.

35 Cited in James C. Ungureanu, *Science, Religion, and the Protestant Tradition*, 4, as a prime example of the position that Ungureanu's book critiques.

36 In a Latin text published in 1681, the *Methodus Theologiæ Christianæ* (Method of Christian Theology), Richard Baxter promised atonement to all Christians, Roman Catholic or Protestant, who had faith and were penitent. His views, which stressed human agency and departed in crucial respects from Calvinism, were promulgated in an English text in 1675 and caused a split among dissenters (see https://crosstheology.wordpress.com/richard-baxter-on-the-atonement/). For the Latin text, see https://archive.org/details/bim_early-english-books-1641-1700_methodus-theologiae-chr_baxter-richard_1681/page/n507/mode/2uBiks.

37 Cited from the 1682 text of *An Excellent Discourse*, as translated by "W.L.," Sig A2v, Baxter's italics. The book does not mention Baxter on the title page but includes his lengthy preface signed at the end.

38 Baxter, *An Excellent Discourse*, sig. A2v.

39 Cited from Vesalius, *The Fabric of the Human Body: An Annotated Translation of the 1543 and 1555 Editions*, ed. Garrison and Hast, Appendix B, Book 5 chap. 15, "On the Uterus and other Female Organs Serving Reproduction," 1084.

40 Henry Newman, "A seale of Virgin waxe at hand," 96. On the importance of hypotheticals in *As You Like It*, see Wofford, "Hymen and the Gods on Stage," and Katrin Ettenhuber, "Hypotheticals and Conditionals," in *The Logical Renaissance*, esp. 249–61. See also Maura Slattery Kuhn, "Much Virtue in If," 46; she argues that Hymen's verse in the play consists of "unprovable hypotheses."

41 On the complexities of Renaissance theories of touch, see Moshenska, *Feeling Pleasures*, and Elizabeth D. Harvey, ed., *Sensible Touch*, especially her "Introduction," 1–29.

42 See A. H. Compier, "Rhazes in the Renaissance of Vesalius," n. 28.

43 See *Vesalius: The China Root Epistle: A New Translation and Critical Edition*, ed. and trans. D. H. Garrison, with a valuable introduction. Garrison bases his translation on a copy of the first edition of the *Epistle* (Basel, 1646), now held in the Bibliothèque interuniversitaire Santé at the Université Paris Cité (formerly Descartes). I cite the digital copy of this volume (as "L") from https://archive.

org/details/BIUSante_00154. My English quotations are drawn mainly from the partial translation by C. D. O'Malley in *Andreas Vesalius of Brussels*, hereafter cited parenthetically in the text by page number. When I use Garrison's version, I refer to it as "G" with page number.

44 See Von Staden, "The Discovery of the Body," 225.

45 For an analysis of mimesis in early modern religious, scientific, and literary discourses, see Kitch, *Shakespeare's Theater of Nature.*

46 See the paraphrase and citation of Thomas Kuhn mentioned earlier and discussed in the Editors' introduction of this volume, 3.

47 In his discussions of Vesalius's early dissections of corpses that supposedly had hymens (63, 201), O'Malley does not translate the story of Vesalius's first dissection (of the six-year-old girl) fully or in its context, so I use Garrison's version for this passage (167, from the Latin text p. 139).

48 Vesalius, *China Root Epistle*, cited in O'Malley, 63 (see also G 169, L 142).

49 Sissa, *Greek Virginity*, 2.

50 See Ilza Veith, *Hysteria,* for an orientation to pre-Freudian theories of this (set of) disorders—usually seen in female virgins but famously transposed to his own body by Shakespeare's *King Lear* (2.4.52).

51 Vesalius, *Fabrica*, cited in O'Malley, 63.

52 Rhazes mentions this remedy in his section on therapies in book 9 of his *Liber de Almansorem* (*c.* 900 CE); Vesalius's first publication was a Latin "periphrasis" of this book (1537). On Vesalius's growing interest in Rhazes as an alternative to Galen, see Compier, "Rhazes in the Renaissance of Andreas Vesalius."

53 Vesalius, *Fabrica*, cited in O'Malley, 63 (see also Garrison, 169–70, L 142).

54 See Loughlin, 43–5, on how the signs of masturbatory desire on the dead girl's part frustrate the frame of objectivity sketched by the anatomist. See also Valerie Traub, "Gendering Mortality," 44–92.

55 See Traub, "The Nature of Norms," 41–82.

56 On the censorship controversy, see Sawday, *Body Emblazoned*, 224–26. O'Malley discusses Crooke's false claim to be authorized by the King in "Helkiah Crooke." On Crooke's ability to move fluidly from allegorical to anatomical discourse, see Elizabeth D. Harvey, "The Touching Organ." For a discussion of the title page image of two humans partially opened by anatomists, with the woman standing in a "Venus Pudica" pose above an inscription from Virgil's *Georgics*, see Newman, *Fashioning Femininity*, 1–2, 11–12. Note that there is no fetus pictured in the woman's womb; the image reproduces a page in the text that contains small letter-labels, one saying that the part portrayed is a womb. Newman observes that the female figure stands on the earth with a bay leaf peeping out; it alludes not to her potential for motherhood but to Crooke's indebtedness to Laurentius as his intellectual progenitor, called a "matrix." On the complexities of Crooke's understanding of gender difference, see Jess Pfeffer, "Trans Materiality."

57 Crooke, *Mikrokosmographia* 197.

58 See Soranus's *Gynecology*, trans. Temkin, 15; for the Greek text and a French translation and commentary, see *Soranos D'Ephèse, Maladies des Femmes*, ed. and trans. Burgière, Gourevitch, Malinas, 1:15.

59 See *Gynaikeia*, Temkin, 7, and Burgière, 7; Soranus yokes such freedom (*adeisdaemona*) with disbelief in "vulgar rites" and "omens."

60 See Augustine, *City of God* 1:18, trans. McCracken. The passage is quoted by Michel de Montaigne in his *Essays* in the translation by John Florio (1603, 3:521). Augustine believes in the hymen's existence as a somatic sign ("integritatem," he calls it) but not in its legibility as testimony to an invisible quality of spiritual "integrity." On Ambrose as another early Christian who reifies the somatic hymen but denies its truth-value, see Sissa, *Greek Virginity*, 172–5.

61 Soranus, *Gynaikeia* (1:7), "Whether Permanent Virginity is Healthful," trans. Temkin, 27–30. Peter Brown takes Soranus to be referring to the Roman Vestal Virgins here, but that seems unnecessarily restrictive given the many early Christians who preferred abstinence to marriage; see Brown, *The Body and Society*, 8–9, 19.

62 Kelly, *Performing Virginity*, 26. She discusses medieval Arabic accounts, rabbinical ones (concerned with the sign of blood, not with a membrane), and modern scholarly debates about the transmission and reception of Soranus's *Gynaikeia*.

63 Crooke, Book 4, Controversy 13, 255. The biblical "testimonies" have to do with the passage in Deuteronomy 22:13-21 about a new husband's demand for "signs of virginity" on sheets or garments. There is no evidence that the ancient Israelites knew of a hymeneal "membrane."

64 See Soranus, *Gynaikeia*, trans. Temkin, 15, in comparison to Crooke, *Mikrokosmographia* 256. On the complexities of defining a corporeal "part," see Hillman and Carla Mazzio, *The Body in Parts*, Introduction and especially xv. The polymath and Royal Society Member Robert Hooke used the new technology of the microscope to define a tiny body part he called a "cell." His concept of a body "part" did not require that it be located in the same place in similarly categorized creatures; instead, the cell was a structure fundamental to all living creatures. After Hooke discussed the cell in his *Micrographia* of 1665, debates about the visible and invisible parts of bodies changed significantly from the debates limned by observers such as Soranus and Crooke.

65 Cited in Peterson, "Elizabeth I's Virginity and the Body of Evidence" (840). She also discusses the possibility that the Queen and her advisors might have planned for a French surgeon (possibly Ambroise Paré) to operate on her before her (possible) marriage to the Duke of Alençon in the mid 1580s.

66 Protestant theologians disputed the Roman Catholic doctrine of Mary's "perpetual" virginity. See, e.g., Calvin, "Commentary on Luke 1:34,"

in *Commentary*, 1:107. He finds the doctrine insulting to the honor of marriage and to the idea of a wife's duty to her husband: a key element of ancient Christianity that is explored by Michel Foucault in *Confessions of the Flesh*, 193–220. Mary's perpetual virginity is arguably important for Elizabeth I's self-fashioning; see Peter McClure and Robin Headlam Wells, "Elizabeth I as a Second Virgin Mary."

67 In *Disknowledge*, Katherine Eggert argues that early modern alchemical science "reflects not an unconscious and inevitable misogyny" but instead a "*willed* ignorance" generated, in part, by practices of tailoring sources "to suit the local needs of author and audience" (168, 170). One might compare such practices with the socio-economically inflected ones that scholars have termed "agnotology," the discourse (logos) of ignorance—as in materials produced by tobacco companies in the U.S. Eggert claims that many famous Renaissance anatomists deliberately held female reproductive organs such as the ovaries in a "kind of half light," as interior phenomena that "could be seen and not-seen" (171).

68 See Sissa, "The Hymen Is a Problem, Still."

69 Sissa, *Greek Virginity*, 2. See also her "Subtle Bodies" and "Maidenhood without Maidenhead."

70 Harrison, "Introduction" to *A Cambridge Companion to Science and Religion*, 15.

3

Hamlet's Soul: Science, Religion, and Theatrical Subjectivity

Aaron Kitch

Ben Jonson's exuberant praise of Shakespeare as the "Soul of the age!" in his dedicatory poem to the First Folio of 1623 also names a particular obsession of his friend and rival. Shakespeare uses the word "soul" and its cognates more than 500 times in his poems and plays, with some forty references in *Hamlet* alone. Such statistics suggest that the playwright alluded to the soul more than any of his fellow playwrights—indeed, more than almost any other published author in early modern England.[1] As this chapter will explore, Shakespeare's interest in the soul as both an organic and ineffable force emerges from centuries of debate in theology and natural philosophy. Early modern authors, including Jean Calvin, Martin Luther, Andreas Vesalius, Philip Melanchthon, and Francis Bacon, revived interest in the soul. Their investigations coalesced into a new field of study known as the science of the soul (*scientia de anima*) that focused on the soul's relation to the body, its immortality, and its shaping influence on human nature.[2] This chapter approaches the soul in *Hamlet* in relation to this new "science," which was simultaneously theological and philosophical in regarding the soul as both material and ineffable. Beyond serving as a locus of suffering and doubt that causes Hamlet to hesitate in revenging the death of his father, the soul is a source of embodied subjectivity that raises central questions about epistemology, matter, and theatricality. Indeed, the soul-spirit dyad that emerges in relation to the ghost fractures Hamlet's sense of self and introduces a paradox of mimesis: even as Hamlet proclaims his deep skepticism toward mere appearances without substance, he also defends the capacity of artificial representations such as painting and theater to reveal truth.

One challenge in representing the soul for artists, theologians, and natural philosophers alike was its troubling invisibility. Even if, as Wittgenstein famously remarked, the body is the "best picture of the soul," such a picture could only reveal itself indirectly and imperfectly since the soul itself remained stubbornly hidden.[3] But the theater of Shakespeare and his contemporaries helped to bring the soul into view in new ways. Notably absent from most works of medieval drama, the soul was addressed more frequently in London plays of the late sixteenth century beginning with Marlowe. Shakespeare expanded on Marlowe's interest in the soul in his own approach to the soul as a source of embodiment, cognition, desire, and subjectivity.[4] Drawing on debates in classical, medieval, and early modern theology and natural philosophy, Shakespeare defines the soul in relation to the mimetic power of theater itself, especially in *Hamlet,* where its identity emerges both in relation to the ghost and to the play's repeated exploration of playacting. In short, the soul was increasingly understood as a site of individual identity during Shakespeare's lifetime, but such identity remained contested and problematic.[5]

Following Donovan Sherman's claim that the soul is "a spiritual form that occupies the contact point between knowing and not-knowing" in the play, I suggest that it is also a vexed source of cognition, embodiment, and selfhood.[6] Rhodri Lewis also links the soul in *Hamlet* to "the animating force and characteristics that make an individual who he or she is," a process that he associates specifically with Cicero's *ingenium*, meaning "how one's individual abilities and disposition [...] allow one best to play a role within the clearly defined arena of public life." For Lewis, Hamlet's insistence on his emotional authenticity as a sufferer—in distinction to those who merely "seem" to suffer—establishes his credentials as a Ciceronian self-analyst, but his rejection of the embodied and emotional soul reduces humanity to a more abstract and cognitive existence, freed from the emotional turmoil of the vegetative soul. Lewis also overlooks important debates in theology and natural philosophy in his account.[7] In my reading, Shakespeare's representation of the soul in *Hamlet* registers a historical tension between the soul as a principle of ontology and of epistemology—of being and of knowledge.[8] And, like the soul, the related concept of spirit invoked by the ghost connotes distinctively material qualities in early modern medicine and alchemy.[9]

Tracing the Soul in Natural Philosophy and Christian Theology

According to Aristotle, it was Thales who first elevated the soul as an object of scientific study in classical Greece, describing it as an invisible force akin to magnetism (405a). Other pre-Socratic philosophers understood at least three types of soul: the disembodied soul of human breath (*psychē*), the embodied soul (*thymos*), and the departing soul of the dead.[10] Although Pindar in his second Olympiad (*c.* 476 BCE) associates the soul with salvation, it was mostly a metaphysical category in ancient Greece.[11] For example, Plato invokes the invisibility of the soul to help make a case for its immortality in *Timaeus* (37a, *c.* 360 BCE) and develops a tripartite model of the soul in *The Republic* (*c.* 370 BCE) using the analogy of a chariot pulled by one black and one white horse. The white horse figures the spiritual soul, which leads the rational soul (the charioteer) in its pursuit of truth and reason. The dark horse represents the appetite and its concern with fulfilling bodily needs such as food and love. The charioteer struggles to keep both horses aligned through space and time.

Despite its foundational role in ontology for both Plato and Aristotle, the soul remains an epistemological problem registered in and through challenges of representation. For example, Plato's tripartite allegory invokes modes of mimetic representation that he elsewhere rejects. Those seduced by art, music, drama, and poetry in *The Republic* elevate their passions over their reason in the same way that tyrants seduce and subject a willing populace (574e-575a). Plato also acknowledges the mimetic challenge of detecting the soul, since it can only be understood in terms of what it "resembles," including images and words (246a).[12] Revising Plato, Aristotle asserted the soul's centrality to ordinary bodily functions and devised an evolutionary model of Plato's tripartite soul that moves up the chain of being from lowly matter to elevated spirit. The vegetative soul thus motivates growth, nutrition, and generation; the organic soul governs bodily sensation; and the rational or "intellective" soul (which was exclusively the provenance of humans) organizes understanding and will.[13] Like Plato, however, Aristotle also figures the soul as a problem of representation. In the opening paragraph of his work *On Interpretation*, for example, he argues that spoken sounds are "symbols of affections in the soul" while written marks are "symbols" of those spoken sounds. Both written and verbal signs are "affections of the soul" that can only be understood in relation to the "likenesses of actual things" (16a).

In a number of works, including *Quod animi mores corporis temperatura sequantur* (That the Capacities of the Soul Depend on the Mixtures of the Body, *c.* 200 CE), the Greek polymath Galen developed a more extensive account of the soul's relation to the body as part of his focus on bodily humors. He defined the soul primarily as an animating principle of biological existence (*bios*) responsible for physical movement, eating, and sexual reproduction. Accepting Aristotle's organic theory of the soul, Galen nevertheless resituated the soul from the heart, where Aristotle had placed it, to the heart, brain, and liver, following Plato's tripartite division. This redistribution linked soul and body even more fully and assimilated the former to Galen's theory of humors.

For Galen, the human body is a complex matrix of ebbs and flows, some visible and others out of sight—what Gail Kern Paster describes as "corporeal flux and openness to the world."[14] The four humors fluctuate constantly in the body and shape three distinct but overlapping forms of spirit: natural, vital, and animal. Galen assumed that blood flowed through invisible capillaries in the veins—a theory that later natural philosophers confirmed with the aid of microscopes. At key stops in the transformational journey from natural to vital to animal spirits, organs of the body such as the lungs mix tangible and intangible substances—blood with *pneuma*, for example. The soul's capacity for traits such as courage or desire is thus rooted in the body, yet the body's own fluctuations indicate that the soul's origin and operation can never be pinned down. It is, rather, fluid and rhizomatic, always evading definition. Because it cannot be comprehended fully and because of its combination of material and immaterial qualities, the soul was also linked to the revival of apophatic theology and forms of religious skepticism in early modern Europe.[15]

Medieval and early modern authors inherited and adapted these classical accounts. Christian theologians struggled in particular to produce a logically and spiritually consistent narratives of the soul. Both Old and New Testaments appear to favor the embodied soul in the manner of Aristotle and pre-Socratic philosophers, even as Plato's account of the soul found new life in Augustine's efforts to defend its immortality. The Hebrew word for soul, "*nepesh*" (נֶפֶשׁ), literally means breath (or, metonymically, the throat through which breath flows). The act of human creation in Genesis, therefore, describes God blending his breath with clay (2:7). Leviticus also locates *nepesh* in the flesh and blood of mankind as a force of vitality (Lev. 17:11), while Psalm 42 suggests, in the King James translation, that the soul is a location of inner desire: "As the hart panteth after the water brooks, so panteth my soul [*nepesh*] for thee, O God." In using the Greek word *psychē* (ψυχή) for "soul," the New Testament generated new accounts of the soul's spiritual transcendence, as in the claim in Revelation 6:9 to witness "souls" (*psychai*) after death or in Matthew 10:28 (King James Version): "And fear not them which kill the body, but are not able to kill the soul [*psychē*]."

Christian efforts to define the soul's immortality often floundered on the sands of logic. In Book 22 of *City of God Against the Pagans* (413-26 CE), for example, Augustine suggests that because God originally endowed humans with souls, he could also reanimate both body and spirit. After death, God turns the "animal body [...] into the newness of the spiritual body," which is "clothed in incorruption and immortality."[16] Where Plato had regarded the soul as the ruler of the physical body, Augustine insists on the unity of soul and body, though he also notes that postlapsarian souls are enslaved to bodily passion.[17] But Augustine sought in vain to clarify ambiguous scriptural passages, including 1 Corinthians, which alludes to the paradox that a human body that is "sown in corruption" could also be "raised in incorruption" at the Last Judgement (15:42). 1 Corinthians solves the problem of how the soul can be raised with the body by suggesting that all people have a "natural body" that only becomes a "spiritual body" through Christ's extension of everlasting life (15:44). A later verse notes in the King James translation that "the first man Adam was made a living soul; the last Adam was made a quickening spirit" (15:45). This fallen man is "earthy" while the saved man is a "Lord from heaven" (15:47). This final "spirit" is the product of the miracle of Christ's salvation, which transforms mortal life into immortality just as it transforms corruption into incorruption (15:53). Such a transformation also depends on forms of mimesis: "As we have born the image of the earthy, we shall also bear the image of the heavenly" (15:49).

In his effort to produce an account of the soul's immortality, Augustine thus invokes without resolving the epistemological and ontological conflicts of classical accounts. Subsequent efforts to define the immortal soul faced similar challenges, especially in the wake of the new theory of faculty psychology that emerged with Aquinas. Some medieval authors described the soul as the real or "true" body of the Christian, while others drew on new translations of the Arabic works of Avicenna (Ibn Sina) on the soul beginning in the twelfth century.[18] Avicenna's *Metaphysics* (*c.* 1014–20) and *Compendium of the Soul* (*c.* 1030) were both steeped in Aristotle, and sometimes Plato, as were the influential commentaries of Averroes (Ibn Rushd) on Aristotle. Where Avicenna endorsed the immortality of the soul as an entity free from the body, Averroes challenged the immortality of the soul by arguing that the soul is inactive after death and therefore essentially a function of the body.[19] These and other medieval works also imagined new forms of representation in connection with the soul. For example, Latin translations of the encyclopedic *Book of Healing* (1027) by Avicenna employed forms of the Latin word *repraesentatio* in the place of numerous Arabic terms to describe mental images and concepts.[20] This term dovetailed with a European understanding of *repraesentatio* as both the act of bringing an idea into the mind and as an artificially produced image or object, a new version of the mimetic structure earlier philosophers had attached to the soul.

Faculty psychology centered on the doctrine of *species* that attempted to account for corporeal forms in nature. For Roger Bacon, such *species*, including light and heat, were simultaneously material and immaterial—invisible to the naked eye but registered in material ways. In *De Multiplicatione Specierum* (On the Multiplication of Species, *c.* 1261), Bacon argues that both the rational and sensitive soul need this quality of receptivity because they must be able to respond to *species* in a physical manner.[21] Aquinas also invoked species in his critique of Averroes and others, defining them as forms of objects that could only be registered by the human mind in terms of spiritual knowledge and being. Both authors worried that demonic forces could corrupt the soul in and through the body.[22] This new field of faculty psychology sought both to identify different mental functions based on their location in the brain and to understand the different "faculties" that organs such as the brain possessed.[23] Faculty psychology achieved its greatest dominance in Europe during the sixteenth century, becoming an empirical science that offered a range of new theories about the physical embodiment of the soul. The most influential work in this nascent discipline was the *Pearl of Philosophy* (1503) by Gregor Reisch. Adapting both Aquinas and Aristotle, Reisch attempted to produce a more empirical account of the soul and its relation to the body. He justified his more materialistic approach on the basis of Aristotle's well-known dictum that all knowledge derives from the senses while also acknowledging the epistemological challenge of viewing the soul "because spiritual (i.e., immaterial) substances, including the soul that confers life and motion on living beings, cannot be perceived by the senses." For Reisch, as for Aristotle, "it is difficult to arrive at knowledge of [the soul] except through its operations."[24] The body for Reisch is a material register of the soul as when, for example, we tilt our heads backward in an effort to recall a name or event—an action that moves fluids toward the posterior ventricle of the brain where such a memory was stored. And, following Galen's theory of bodily humors, the human face also registers soulful emotions such as anger by becoming red and swollen.[25]

Reisch provides a materialistic account of the "common sense" (*sensus communis*) that gathered and compared information from the body. He suggests that the medium of that information is images or similitudes that found their way to the Imagination, which in turn allowed Fancy or "fantasy" to reassemble them into new images or *phantasmata* that capture and even transform sense impressions. Over time, works of faculty psychology developed a materialistic and representational theory of the operation of the soul within the body that centered on *spiritus*, a substance derived from blood that travelled between organs of the body as a life essence. In his treatise on *The Laymen on Mind* (1450) and his later *Compendium* (1463), for example, Nicolas Cusanus followed Reisch in arguing that all created matter generates invisible "species" that take the form of *spiritus*, which can include images. Such species are both material and immaterial, since they constitute types of natural signs that interact with the mind in and through the *spiritus*. It is only when the mind or "soul" is properly situated within the *spiritus* that it can be prepared to receive sensual data. For non-sensible things, the mind can itself produce a mode of "auto-excitation."[26]

As these examples suggest, early modern Europe inherited a complex and sometimes contradictory set of ideas about the soul as both embodied and spiritual, both organic and invisible. Leonardo da Vinci, for example, focused not on the spiritual status of the soul but on its location in the body. Following Aristotle, he situated it in the *sensus communis* of the human brain and focused on the soul primarily as an agent of human movement and sensation derived from nature and light that enables both observation, sensation, and aesthetic awareness.[27] Thus the soul for da Vinci creates "the most complete and magnificent view of the works of nature."[28] Good paintings achieve a type of "grace" that "plays with the soul," while those who wish to learn how to fly "must counterfeit the soul of the bird" in addition to its body.[29] Like many of the other writers considered above, da Vinci mixes Aristotelian and Platonic traditions of natural philosophy in his account of the soul as foundationally embodied and yet transcendent—and as intimately associated with forms of representation. The soul defines humanity for da Vinci not because of its spiritual immortality but because of its active role in organic sensation, cognition, and aesthetic enjoyment. By linking the soul primarily with the capacity of the body to perceive the order of nature, da Vinci and others see it as a window to the outside world. Fundamentally embodied, the soul is also valued precisely because of its capacity to transcend the body. In foregrounding the soul's relation to vision, moreover, da Vinci anticipates Descartes and others in the seventeenth century who linked the soul to new kinds of subjectivity. As the soul became more firmly aligned with the mind in early modern Europe, in other words, it became valued as an immaterial form of consciousness that was the basis of individual identity, even while it remained thoroughly enmeshed with the body.

Descartes not only worked to align soul with mind as separate from the body; he also aimed to restrict the soul that Aristotle had located in animals and vegetables exclusively to humans. He and other mechanical philosophers eventually identified mind rather than soul as the seat of reason, human identity, and immortality. In the context of these shifting arguments about the soul in both scientific and religious texts, it is no wonder that Shakespeare's contemporary John Davies was moved around the same year that Shakespeare began writing *Hamlet* to satirize the confused efforts of theologians and natural philosophers to locate the soul:

> Some place it in the root of life, the *Heart,*
> Some in the *Liver,* fountain of the Veins
> Some say, *she is all in all, and all in part*:

Some say, she is not contained, but all contains.
Thus these great Clerks their little wisdom show,
While with their Doctrines they at *Hazard* play,
Tossing their light opinions to and fro,
To mock the *Lewd,* as learned in this as they.[30]

Davies satirizes the numerous and contradictory "opinions" of the "learned" concerning the soul, registering the epistemological and spiritual confusions of the soul as the foundational but epistemologically troubling "root of life." Shakespeare's theatrical framing of the soul shares some of Davies's humor while also exploring further questions about the soul as a source of subjectivity and mimesis.

Hamlet and the Staging of the Soul

My survey of printed works of drama in England before 1580 reveals only around sixty references to "soul" and its cognates, including the Plowman in John Rastell's *Of Gentleness and Nobility* (1525) who claims that "man is the most noble of creatures living" because of his "soul" (and especially his "soul intellective") that allows him to conquer animals.[31] But playwrights beginning with Marlowe invoked the soul more frequently, including a passage in *Dr. Faustus* where Faustus, facing imminent death, wishes he were born without a soul and then hopes to inhabit new bodies according to Pythagorean metempsychosis.[32] In *Tamburlaine* (*c.* 1587), moreover, Marlowe refers to the soul as a "faculty" that is part of the divine Book of Nature. As Tamburlaine explains, assuming the stance of a natural philosopher,

Nature that framed us of four elements,
Warring within our breasts for regiment,
Doth teach us all to have aspiring minds:
Our souls, whose faculties can comprehend
The wondrous Architecture of the world:
And measure every wandering planet's course
Still climbing after knowledge infinite,
And always moving as the restless Spheres.[33]

Tamburlaine defines the soul as a material "faculty" that has been embedded in us by God through the divine agency of Nature. This faculty fosters human comprehension of the "wondrous Architecture of the world," even though its capacity to discern that architecture and to climb after infinite knowledge is compromised by the warring elements within us. Yet Marlowe importantly inserts the soul into the pursuit of natural knowledge that Tamburlaine praises. It is the soul that provides us with the power to observe the Book of Nature, which was also a site of empirical forms of natural philosophy in the sixteenth century.[34] In defining the soul as a source of curiosity, desire, and ambition, moreover, Marlowe reveals the centrality of the soul to the construction of his central character.

Shakespeare extends Marlowe's interest in the soul as a product of the divine order of nature and as a problem of cognition, embodiment, knowledge, and theatrical mimesis. At times in his plays, Shakespeare echoes Aristotle's definition of the soul as an organic life force, as in the Duke of Exeter's reference to the "poor souls for whom this hungry war / Opens his vasty jaws" in *Henry V*.[35] The soul can also take on Christian meanings, as when the doomed Clarence attempts to reason with the murderers hired to kill him, asking them "art thou yet to thine own soul so blind / That thou wilt war with God by murd'ring me?" (1.4.232-4). But Shakespeare's souls are not exclusively Christian—nor even exclusively human.[36] In *The Merchant of Venice,* for example, Graziano invokes the metempsychosis of Pythagoras ("that souls of animals infuse themselves into the trunks of men") in order to joke that the "currish spirit" Shylock is in possession of the "soul" of a wolf (4.1.131-4). And Martius (before he becomes Coriolanus) in *Coriolanus* insults retreating soldiers by saying that they have "souls of geese" (1.4.35).

Souls also serve as proxies for emotional depth in Shakespeare. For example, the newly crowned King Henry IV notes in *Richard II* that his soul is "full of woe" after the killing of Richard II, since the former king's "blood" is what must "sprinkle" him to "make" him "grow" (5.6.45-6). And when Bolingbroke accuses Mowbray of treason in the opening scene of *Richard II*, he vows that "my body shall make good upon this earth / Or my divine soul answer it in heaven" (1.1.37-8). What he means by this is that he will put his body forward in direct combat with Mowbray to either prove the truth of his accusation or die in the process, where his soul will bear witness to his life. In another historical tragedy, following the murder of Caesar in *Julius Caesar*, Mark Antony displays the slain body of the general, imploring the multitude of "kind souls" about him to "weep" and feel pity (3.2.193).

Like the tears of the plebeians in *Julius Caesar*, the soul in Shakespeare is almost always embodied. Near the end of *Coriolanus*, for example, Menenius Agrippa volunteers to beg Coriolanus not to sack Rome. Where an earlier mission by Cominius had failed, Menenius vows success because he will ask

Coriolanus after the general has had his supper, a time when the blood is "stuffed" with "wine and feeding," generating "suppler souls" (5.1.53-54). Menenius seems to allude here to the Galenic idea that digestion feeds the soul, although he does not get to test his theory because Coriolanus refuses to meet with him. Elsewhere, Shakespeare stages the soul in relation to problems of visual and verbal representation. As she confirms her love for Othello in the face of skeptical questioning by Senators of Venice in the first act of *Othello*, for example, Desdemona claims to have seen his "visage in his mind" and vows to "consecrate" her "soul and fortunes" to his "honor and valiant parts" (1.3.249-51). These wide-ranging examples of the soul in Shakespeare raise questions about embodiment, theology, and representation that are linked both to questions of character identity and to the capacity of the stage to represent invisible forces. And nowhere in Shakespeare is the soul more central to questions of subjectivity, science, and theology than in *Hamlet*.

Critics since the nineteenth century have celebrated Hamlet's psychological depth and complexity. William Hazlitt, for example, described Hamlet's ideas "as real as our own thoughts."[37] Later interpretations challenged such accounts from a number of perspectives. New historicists and cultural materialists writing at the end of the twentieth century, for example, suggested that Hamlet's famed interiority was an anachronistic projection of later readers. To Hamlet's claim early in the play he has "that within which passeth show," they responded, in essence, "no you don't." As John Lee notes in his study of the "controversies of the self" surrounding the tragedy, "a vocabulary of essentialist interiority cannot be found within *Hamlet* because it does not exist outside it; its absence in *Hamlet* is a reflection of its absence from the wider culture."[38] This claim echoes those of previous scholars that Hamlet's apparent interiority depends on outward-facing public and rhetorical structures. Thus Stephen Greenblatt, who mentions Hamlet only once in his foundational study of *Renaissance Self-Fashioning*, links his "inner life" to public forms of rhetoric such as the *quaestio*.[39] Frances Barker likewise defines the "metaphysics of interiority" apparent in Hamlet as a mode of "bourgeois subjectivity" that emerged only in later decades.[40] These critical accounts, shaped by anti-humanist theories of post-Freudian psychoanalysis, Derridean deconstruction, and Foucault's theory of power, overlook a more relevant historical locus of questions of the self for Shakespeare and his contemporaries: the soul.

The soul is a source of embodied if conflicted subjectivity for multiple characters in *Hamlet*. Shakespeare invokes it to describe the source of speech in Polonius (1.3.115-16) and of Hamlet's own distracted thoughts (1.4.56), for example. But the play initially frames the soul in relation to the spirit of the ghost of Hamlet's father who appears first in 1.2 as Horatio arrives at court to tell Hamlet that he has seen the specter on the ramparts of Elsinore. In response to Horatio's strange and unsettling tale, Hamlet reflects on his own spiritual state:

> My father's spirit in arms! All is not well:
> I doubt [i.e., suspect] some foul play. Would the night were come!
> Till then sit still, my soul. Foul deeds will rise,
> Though all the earth o'erwhelm them, to men's eyes.
> (1.2.254-57)

Fearing that his father has been murdered, Hamlet cautions his own private soul to "sit still" and not avenge the death until the hidden "foul deeds" "rise" and overwhelm the "eyes" of men. The external "spirit" of the ghost triggers this intense scrutiny of his own soul, a process repeated when the Ghost finally speaks to Hamlet, announcing "I am thy father's spirit" (1.5.9) and promises to reveal a story that will "harrow up thy soul" (line 16). Such a "soul" is a chief agent of revenge against Claudius, yet both Hamlet's soul and spirit remain stubbornly inaccessible to the audience—and even to Hamlet himself. Both also stand metonymically for Hamlet's own effort to grieve for his father, an effort itself framed by larger problems of representation. As he says to his mother in 1.2, his black suits of woe indicate "that within which passes show" (1.2.85).[41]

As Abe Davies notes, the ghost of Hamlet's father differs from other stage ghosts in early modern drama by virtue of its materiality.[42] The ghost first appears in a suit of armor with his beaver up, a mysterious specter who will speak only to Hamlet and is visible to some characters but not others. He is neither a charmed supernatural agent like Puck nor a committed agent of revenge in the Senecan tradition. Rather, he is a radically ambiguous figure who journeys from the Catholic realm of Purgatory to disrupt the political and epistemological order of the "world" of Denmark, causing what Horatio calls "some strange eruption to our state" (1.1.68). His very presence in the play, in fact, may owe something to a folk tradition captured in medieval debates between the body and soul in which souls of sinners are periodically released from their confinement in the afterworld to haunt the space of their death at night until the cock crows.[43] The ghost first appears as a figure "like the King" (line 57), although Bernardo, Marcellus, and Horatio speak of it as a "guilty thing" (line 147) and a "present object" (line 155). Such metaphysical uncertainty prompts Horatio and Marcellus to propose contrasting theories of its origin. Invoking Greek mythology, Horatio suggests that the "shrill-sounding" cock wakes the "god of day," who forces the "erring spirit" into his place of confinement (lines 148–54). Marcellus offers an alternative interpretation that invokes Christianity for the first time in the play, albeit one mixed with supernatural beliefs and folk knowledge.

> It faded on the crowing of the cock.
> Some say that ever 'gainst that season comes
> Wherein our Savior's birth is celebrated,
> This bird of dawning singeth all night long,
> And then, they say, no spirit dare stir abroad,
> The nights are wholesome, then no planets strike,
> No fairy takes, nor witch hath power to charm,
> So hallowed and so gracious is the time.
> (lines 156–63)

Marcellus evokes a conventional narrative of Christ's birth silencing the oracles—a moment when the sudden intrusion of divine logos into the everyday world strips witches, fairies, and planets of their supernatural powers. The ghost confirms in his absence what his presence in *Hamlet* suggests—that he stands for a world of astrological influence and the dangerous charm of witchcraft.

Though largely associated with the ghost in the first part of the play, "spirit" becomes more of an attribute of individual agency as the play unfolds. Of the 19 allusions to "spirit" and "spirits," the first eleven refer to the ghost but none of the final eight do. These later instances largely concern elements of subjectivity. In act 3, for example, Hamlet praises Horatio's "good spirits" (3.2.51) in his refusal to be "passion's slave" (line 65). In the *Murder of Gonzago*, moreover, the Player King complains that his "spirits grow dull" as he prepares to sleep (line 210). Guildenstern then describes Gertrude's "great affliction of spirit" after she retires to her closet following the aborted performance of the Mousetrap (lines 288–9), while Hamlet pictures the "spirit" of Fortinbras "with divine ambition puffed" (4.1.48).

Hamlet's own celebrated interiority unfolds in and through this dialectical dance of soul and spirit, which also helps to account for the play's relentless materialism. Some critics have understood such materialism in theological terms. For Greenblatt, for example, Hamlet's fixation on "too too solid flesh" reflects the barrenness of a culture exhausted by the Catholic claims of transubstantiation.[44] The soul connects Hamlet with passions he may not be able to control and with the destiny of the state to which he is heir apparent. Because it cannot be seen or touched, moreover, the soul joins a larger pattern of revealing and concealing in the play, especially in efforts to discern the psychological "matter" with Hamlet. The soul is both a potential and material source of Hamlet's melancholy, according to Claudius (3.1.161-2), even though elsewhere in the play it appears more metaphysical. In his interactions with the ghost, moreover, Hamlet regards both his father's spirit and his own soul as immaterial, claiming to value his life but at a "pin's fee," since his "soul" cannot be affected by a "thing immortal" such as the ghost itself (1.4.65-7). Yet in claiming that his tale of murder at the hands of Claudius will "harrow up" Hamlet's "soul," the ghost uses a farming metaphor that compares the soul to a field in need of plowing. According to the ghost, Hamlet will register such harrowing in his eyes, which will be forced to "start from their spheres" (1.5.14-17). For his part, Hamlet wonders at the ghost's appearance, which makes "fools of nature" such as himself generate "thoughts beyond the reaches of our souls" (1.4.54-6). This vacillation between material and immaterial soul and spirit both designates and troubles sites of potential subjectivity—appropriately enough for a tragedy whose central monologue begins with the question "To be or not to be?"

The problem of the soul's legibility also relates to the play's focus on forms of theatrical performance as it takes on new meanings with the arrival of the players in act 2. After observing the First Player's speech on the fall of Troy, which includes an account of the suffering of Hecuba, Hamlet wrestles with his own capacity for action:

> O, what a rogue and peasant slave am I!
> Is it not monstrous that this player here,
> But in a fiction, in a dream of passion,
> Could force his soul so to his own conceit
> That from her working all the visage wanned,
> Tears in his eyes, distraction in his aspect,
> A broken voice, and his whole function suiting
> With forms to his conceit?
>
> (2.2.469-76)

Hamlet acknowledges that theatrical performance allows the invisible, inner soul to discover its proper, external form. A crucial irony emerges, however, since his witnessing of theatrical performance in which interior soul finds proper expression in the body reveals a failure of agency in his own soul. Hamlet repeats this link between the soul and playacting in his instructions to the players before the performance of *The Murder of Gonzago* in 3.2. In urging the players to show restraint in both their speech and action, he observes that "it offends me to the soul to hear a robustious periwig-pated fellow tear a passion to tatters, to very rags, to split the ears of the groundlings, who for the most part are capable of nothing but inexplicable dumb-shows and noise" (3.2.7-11). The soul thus bridges relations between drama and lived experience, as in the following scene when Hamlet praises Horatio for his refusal to put his heart into his mouth, to "let the candied tongue lick absurd pomp / And crook the pregnant hinges of the knee" (3.2.53-4). It is this quality above others that makes Hamlet's "soul" seal Horatio to "herself" (3.2.58).

This theatrical scaffolding of the soul informs a series of experiments in the play that involve both professional and amateur actors. Most obviously, Hamlet hopes that the performance of the Mousetrap will render the soul of Claudius visible, noting that "guilty creatures sitting at a play / Have by the very cunning of the scene / Been struck so to the soul that presently / They have proclaimed their malefactions" (2.2.508-11). He also alludes to his own soul as a capacity for discerning the "election" of others (3.2.56-7) and encourages Horatio to employ the "very comment" of his "soul" in paying close attention to Claudius during the performance (line 72). During the performance of the Mousetrap, moreover, Hamlet reassures Claudius that the play "touches us not" since both have "free souls" (line 224). As Michael Witmore notes, Hamlet's theatrical experiment allows him to function as a Baconian natural philosopher who attempts "to access hidden truths by staging an 'accident' and then observing what happens."[45] This metatheatrical subplot literalizes scenes of observation, interruption, and accidents that define the play from beginning to end, from the sudden appearance of the ghost in the first scene to the fatal accidents of the plot to kill Hamlet in the last scene. The spiritual and material condition of the soul—and Hamlet's soul in particular—is thus one of many objects of scientific inquiry in the play.

On his way to visit Gertrude in her "closet" following the aborted performance, Hamlet vows to be gentle with her, hoping that the "soul of Nero" will not enter his body in order to make him "cruel" (3.2.365-6). Even though he has railed throughout the play against the hypocrisy of courtly life, which forces men and women to put on false faces, he freely confesses that he will not speak from the heart:

> I will speak daggers to her but use none:
> My tongue and soul in this be hypocrites.
> How in my words somever she be shent,
> To give them seals never my soul consent.
> (3.2.367-70)

Hamlet acknowledges his own duplicity in these lines but imagines his soul to be empowered by royal decree to be hypocritical. Such an analogy renders his tongue and his soul at odds with one another. He will restrict himself to using verbal "daggers" even though his soul would prefer to use real ones.

The theatricalization of the soul in the tragedy thus attempts to solve the problem of the soul's invisibility while also raising new problems about the soul as a source of subjectivity. Where other members of the court depend on the deceptive capacity of appearances for their own power—especially Polonius and Claudius in their endless acts of spying, covering up, and attempts to manipulate others—Hamlet cannot transcend the shifting and conflicting relations between appearance and reality that he sees everywhere. In contrast to Hamlet's famous claim to possess "that within which passeth show," the tragedy reveals the opposite—that nothing in the world of Denmark can remain beyond "show."

Like Marlowe, Shakespeare also situates the soul in relation to ideas of nature. In urging his sister to be wary of Hamlet's love, for example, which he describes as a mere "toy in blood" and a "violet in the youth of primy nature" (1.3.6-7), Laertes invokes Hamlet's soul:

> For nature crescent does not grow alone
> In thews and bulks, but as this temple waxes,
> The inward service of the mind and soul
> Grows wide withal. Perhaps he loves you now,
> And now no soil nor cautel [i.e., deception] doth besmirch
> The virtue of his will. But you must fear,
> His greatness weighed, his will is not his own.
> (1.3.11-17)

Although Laertes echoes medieval natural theology and philosophy in regarding the "mind" and "soul" as complementary, the fact that he lists them separately gestures toward a growing separation of the mind from its soulful embodiment in the wake of faculty psychology. The specific analogy that Laertes develops here also imagines the soul as a force of nature that makes the body grow physically in "thews" or muscles that shape the "temple" of the body. According to this logic, Hamlet's apparent desire for Ophelia is necessarily "circumscribed" by the demands of his princely status (1.3.21-2). As heir to the throne, Hamlet must consider the needs of the state in choosing a marriage partner. Laertes thus warns Ophelia to be restrained in her affection, to avoid showing her beauty, even before the "moon" (line 36). He ends his patronizing speech in two additional analogies to nature—that cankerworms injure the young shoots of flowers in the spring (line 38) and that "contagious blastments" (line 41) or blights are most powerful in the morning time of youth. Echoing Aristotle's claims about the organic soul, the soul for Laertes becomes a product of amatory passion that operates according to the principles of natural growth, serving as a foil to Hamlet's approach and anticipating the fatal tension between the two characters.

Polonius repeats this natural language when he subsequently warns Ophelia about Hamlet's desire for her, which he finds dangerous. On hearing of Hamlet's "many tenders / Of his affection" to his daughter, Polonius dismisses her as a "green girl" who is unaware of the danger she faces (1.3.98-100). He describes Hamlet's "vows" of love as mere

> springes to catch woodcocks. I do know
> When the blood burns how prodigal the soul
> Lends the tongue vows; these blazes, daughter,
> Giving more light than heat, extinct in both
> Even in their promise as it is a-making,
> You must not take for fire.
>
> (1.3.114-19)

For Polonius, the soul is an excessive ("prodigal") and untrustworthy source of false love that generates "blazes" or images instead of genuine fire. Proposing a notably different relationship with the order of nature than Laertes, Polonius mentions modes of artifice such as the traps or "springes" that hunters use to catch small birds and the imitation of fire that can be produced by artificial fireworks. In contrast to Hamlet's embrace of the soul as a form of artificial representation, Polonius regards it merely as an embodied instrument of deception. Polonius here associates the soul with the play's prevalent language of hunting and trapping that is both animalistic and spiritual.[46] The same analogy appears again when Hamlet encounters Claudius in prayer on the way to visit his mother. Claudius prays for his own "limèd soul" that struggles to be free of his crime of fratricide (3.3.68). Even though he acknowledges the providential opportunity to complete his revenge against Claudius, Hamlet refuses to kill him because doing so would "purg[e] his soul" while sending Claudius to heaven (line 85). In contrast to his earlier "To be or not to be" monologue, which does not recognize a Christian afterlife, Hamlet here speaks of heaven as a given.

In the closet scene, Hamlet reframes the soul once more as a spiritual entity that raises problems of verbal and visual representation. Employing an image of embodied sacramentality, he accuses his mother of having committed "such a deed / As from the body of contraction [i.e., the marriage contract] plucks / The very soul, and sweet religion makes / A rhapsody of words" (3.4.45-8). To illustrate his claim that Gertrude has desecrated the sacred "soul" of marriage itself by marrying the brother of her dead husband, he forces her to view two images, the painted "counterfeit presentment of two brothers," drawing sharp contrasts between the personage of his father, "the front of Jove himself," and his weaker brother, who appears like a "mildewed ear" (lines 53–64). In most performances of the play, Shakespeare's audience does not see these images directly. We only have his description of them, so we must take them on faith or else register their accuracy in and through the reaction of Gertrude, who herself draws attention to the face of Hamlet as a way of registering the reappearance of the ghost. But Hamlet's use of such images succeeds, he believes, because those images lead his mother to beg him to "speak no more." Gertrude says that she has "turns't my very eyes into my soul" and found "such black and grièved spots / As will not leave their tinct" (lines 88-91). This ambivalent materiality of the soul makes Gertrude regard the visible representations of the two men in both material and spiritual terms, accepting Hamlet's argument that her soul is stained by her apparent adultery. Hamlet here convinces his mother of the necessary intervention of artificial imitation—pictorial rather than theatrical this time—in order to visualize the soul.

It is at this moment that the "spirit" of Hamlet's dead father returns to haunt the play, urging Hamlet to "step between" Gertrude and "her fighting soul" (3.4.112). Yet Gertrude cannot see the specter with her own eyes. She asks Hamlet why he seems to "hold discourse" with the "incorporeal air" (line 117). Hamlet responds by describing the pale skin of the ghost, but Gertrude observes nothing and accuses Hamlet of falling ill to the "very coinage of your brain" (line 138). In response, Hamlet asserts his own embodied existence:

> My pulse as yours doth temperately keep time
> And makes as healthful music. It is not madness
> That I have uttered. Bring me to the test
> And I the matter will reword, which madness
> Would gambol from. Mother, for love of grace,
> Lay not that flattering unction to your soul
> That not your trespass but my madness speaks.
> It will but skin and film the ulcerous place,
> Whilst rank corruption, mining all within,
> Infects unseen.
>
> (3.4.141-50)

Extending earlier associations of the soul with Hamlet's own subjectivity, Hamlet argues that Gertrude's soul designates both her spiritual condition and her capacity to look within her conscience. Hamlet's language of putrefaction and invisible corruption here mixes with natural philosophy and Christian theology. The word "unction" appears only twice in all of Shakespeare—both times in *Hamlet*. Traditionally, the Catholic sacrament of "extreme unction" was administered to men and women before death to prepare them for dying and the afterlife. Theologically, it represented the spiritual power of the Church in material form, drawing on biblical passages in Mark 6:13 and James 5:14-5.[47] It was also a source of debate among medieval and early modern theologians. The Council of Trent, for example, codified a Thomist reading of extreme unction as "the consummation not only of penance but also of the whole Christian life, which ought to be a continual penance." John Wycliffe and later reformers declared the practice to be an extra-biblical invention of Rome. At stake in the debate was the capacity of material substances to serve as a conduit for divine penance and, by extension, the degree to which the soul can be affected by material constructions.[48]

Unction was also an important topic of medieval and early modern natural philosophy in relation to the question of how elements combine to form matter. Islamic alchemists whose work circulated in Europe in Latin translations beginning in the late twelfth century had, for example, identified unctuous moisture in certain metals and rocks that made them cohere together while also allowing them to be both malleable and inflammable.[49] The "flattering unction" that Hamlet accuses his mother of applying to her soul thus becomes a metaphor to describe a medical ointment that may relieve pain without having the capacity to heal her soul.[50] A second reference to "unction" in the play underscores this material meaning in the form of the poisonous "unction" that Laertes will touch to his foil in the fatal duel with Hamlet (4.4.140). Laertes praises the poison for its capacity to defeat any medical remedy drawn from the "virtue" of nature (line 143).

This reference to nature recalls the earlier association of the soul with modes of nature by Laertes and the tradition of invoking the soul in relation to natural philosophy in Marlowe. Shakespeare also echoes this tradition in the following scene when, questioned by Laertes, Claudius refuses to kill Hamlet because Gertrude "lives" by Hamlet's "looks" just as Claudius lives for her: "She is so conjunct to my life and soul / That as the star moves not but in his sphere / I could not but by her" (4.4.14-16). This direct allusion to the heavenly spheres of Ptolemy to describe the soul adds another scientific context to the framing of the soul in the play. In response to Hamlet's rash stabbing of Polonius and fearful of Hamlet's threat to his own power, Claudius exiles the prince to England. On his journey there with Rosencrantz and Guildenstern, Hamlet spies Fortinbras and his army near the Danish coast. In speaking with a Captain of the army, Hamlet integrates the soul in his assessment of the escalating tension between Norway and Poland:

> Two thousand souls and twenty thousand ducats
> Will now debate the question of this straw.
> This is th'impostume of much wealth and peace
> That inward breaks and shows no cause without
> Why the man dies.
>
> (4.1.24-28)

Reviving his obsession with the relation between inner essence and outward appearance, Hamlet incorporates the shapeless "souls" of enemy soldiers into a meditation on the health of the body politic. The odd word "impostume," which appears only three times in Shakespeare's works, designates an abscess that eats away at a body from within without revealing any external "cause" of illness. In poetic terms, "impostume" acts to conceal the soul as a form of organic life from the previous line, illustrating the principle of biopolitical death without visible cause that the passage itself proclaims. Relentlessly focused on relations between interior and exterior, Hamlet is especially conscious of how a nation or body rots invisibly from within, without revealing any external signs of hidden disease. The soul's problematic legibility is even more dangerous when figured as part of a medical metaphor for the ailing and self-consuming body politic.

In conclusion, the soul in *Hamlet* draws on unresolved debates within natural philosophy and theology. As Sherman remarks, the "soul must be understood and yet it must not resolve itself into a pedagogy of meaning and things."[51] Where Polonius, Claudius, and Laertes speak of the soul as an unproblematic ground of individual subjectivity, Hamlet in his focus on his own unrepresentable interiority and fascination with the histrionic mimesis of playacting troubles assumptions about the soul as a source of identity. The soul for Hamlet is immanent rather than transcendent. It moves restlessly between the invisible materiality of its sinful origin in fallen flesh and an external system of signs associated with theatrical performance and image-making that cannot almost by definition succeed in denoting the soul truly. Hamlet is the only character in the play who consciously recognizes that he inhabits this liminal space that "passeth show." His celebrated interiority is in fact a direct product of this paradox that refuses to resolve the soul as either material or disembodied. This complex negotiation and renegotiation of the soul's essence and performativity in Shakespeare resists self-evident claims about the soul as the immortal and ineffable ground of human identity. Following theologians and natural philosophers alike, the soul in *Hamlet* is both spiritual and material—metaphysical but also entwined with the body, as illustrated by the material "spirit" of the ghost. By linking the soul to the play's self-awareness as a work of theater, moreover, Shakespeare's play also participates in broader reconfigurations of science, religion, and modes of mimesis in early modern Europe.[52]

Notes

1 I count 553 references to "soul" and its cognates in Shakespeare's first folio (1623). For context, the Early English Books Online (EEBO) database includes 35,378 instances of "soul" and its cognates in printed works published in England between 1473 and 1700, although this number includes only those works that have been encoded by the Text Creation Partnership. Of these, 6,422 are by "Anonymous," 180 appear in works attributed to Richard Baxter, 151 in those by William Prynne, and 143 by George Fox. The collected published writings of Jean Calvin and William Tyndale each reference the soul about 100 times each, while the edition of Montaigne's *Essays* first published in England in John Florio's translation in 1603 contains 421 references to the soul.

2 For an important historical account of the soul in relation to these disciplines, see Fernando Vidal, *The Sciences of the Soul*. On the relation between anatomy and the soul in early modern Europe, see Richard Sugg's chapter on Vesalius and Pomponazzi in *The Smoke of the Soul*, 51–80.

3 Wittgenstein, *Philosophical Investigations*, 131. Virginia Woolf also noted in her diary that "one can't write directly about the soul" since, when "looked at, it vanishes." See *A Writer's Diary*, 84.

4 There is a long tradition of scholarship on the soul and its relation to the self. As Ralf Haekel notes, "No matter at which point in history we choose to determine the beginning of the modern individual self, it is inextricably linked with the history of the soul." See *The Soul in British Romanticism*, 179. See also Timothy J. Reiss, *Mirages of the Selfe*. For studies of the soul in relation to spiritual traditions, see Richard Sugg, *The Secret History of the Soul* and *The Smoke of the Soul*.

5 The *OED* registers this new sense of the soul as a source of individual identity beginning in the late sixteenth century (def. I.5.a), even as the word retained earlier definitions as an allegorical personification (I.5.b) and an artistic representation (1.3.b).

6 Sherman, *Second Death*, 165.

7 See *Hamlet and the Vision of Darkness*, 109, 21. Katharine Bootle Attié also suggests that Shakespeare draws on a Platonic understanding of the soul as "distended" and "out of balance" in her account of tyranny in *Hamlet*, which she also understands in relation to problems of mimesis. See "Tragic Proportions," esp. 90–91.

8 On this division between having and being, see Sherman, *Second Death*, 3 et passim.

9 According to the *OED*, "spirit" could also mean a "person's nature, character, or temperament" (def. I.ii.8.b), while like "soul" it also retained earlier association with physical sensation and cognition.

10 In Homer's *Odyssey*, for example, the *psychē* (literally "life force") journeys to Hades after death, where it mixes with the mindless souls or *eidola* of the dead (11.476).

11 On Pindar, see Asya C. Sigelman, *Pindar's Poetics of Immortality*, esp. 130–2.

12 For more on the soul in relation to representation in Plato, see Marina Berzins McCoy, *Image and Argument in Plato's Republic*, 100.

13 For a fuller account of the soul and theology in Aristotle, see Eli Diamond, *Mortal Imitations of Divine Life*.

14 Paster, *The Body Embarrassed*, 14.

15 See, e.g., Kevin Killeen, who addresses the soul only in passing in his wide-ranging and valuable study of *The Unknowable in Early Modern Thought*.

16 Augustine, *The City of God*, ed. and trans. Tods, 845. See also François Bovon, "The Soul's Comeback," esp. 389.

17 Augustine, cited in Eugene, 32. See also *The City of God*, in *Nicence and Post-Nicene Fathers*, ed. Schaff, 2:259. On Augustine's account of the soul, see Raymond Martin and John Barresi, *The Rise and Fall of the Soul and Self*, 71–2.

18 See, e.g., Caroline Walker Bynum's chapter on "Somatomorphic Soul and *Visio Dei*" in *The Resurrection of the Body in Western Christianity*, 279–317.

19 For an overview of the debate between Averroes and scholastic philosophy, including the *Sentences* of Peter of Lombard that defended the orthodox position of the immortality of the soul and later attempts to justify the immortality of the soul by Ficino, see Richard Blum, "The Immortality of the Soul," esp. 212–16.

20 On Avicenna and his relation to *repraesentatio* and Western philosophy, see Henrik Lagerlund, "The Terminological and Conceptual Roots of Representation in the Soul in Late Ancient and Medieval Philosophy," esp. 13–24.

21 Bacon, *Perspectiva*, ed. Lindberg, 15–17.

22 For more on this fear on the part of both philosophers, see Amanda Powers, *Roger Bacon and the Defence of Christendom*, 133–4.

23 For an account of the significance of the human senses to faculty psychology in Aquinas, see Anthony J. Liska, *Aquinas's Theory of Perception*.

24 Reisch, *Margarita philosophica*, 409. Quoted by Katharine Park in "The Organic Soul," 466–7.

25 Those who wished to become skilled in discerning the presence of the soul thus had to become careful readers of the body, a principle that motivated Ficino, Burton, and others in describing the symptoms of melancholy as well as those who practiced the art of reading facial gestures or the divinatory art of chiromancy. For Ficino, as Sorana Corneanu notes, "since it is the soul that governs (i.e., fashions, nurtures, and ensures the stability of) the body, in cases of bodily disease the soul leaves aside the preoccupations characteristic to it (the search for truth), attends to the body's needs and becomes involved in its cure." See "The Nature and Care of the Whole Man," esp. 146.

26 On Cusanus and on species more broadly, see Leen Spruit, *Species Intelligibilis*, 23–4.

27 John Henry perhaps overstates the case when he observes that "every divine who sought to expound upon the nature of the immortal soul sank into contradiction and incoherence or glossed over the issues so quickly that too many questions remained unanswered," but the difficulty in defining and defending the soul's immortality persisted. See "The Matter of Souls," esp. 97.

28 Da Vinci, *Notebooks,* trans. Richter, 105.

29 Da Vinci, *Thoughts on Art and Life,* trans. Baring, 75, 118.

30 Davies, *Nosce Teipsum*, 11. See also Suggs's discussion of this passage in *The Smoke of the Soul,* 150.

31 Rastell, *Of Gentilness and Nobility* (lns. 377–8), in *Three Rastell Plays*, ed. Axton, 106.

32 "No end is limited to damnèd souls. / Why wert thou not a creature wanting soul? / Or why is this immortal that thou hast? / O, Pythagoras' *metempsychosis*—were that true, / This soul should fly from me, and I be changed / Into some brutish beast." Marlowe, *Dr. Faustus* (B-text, 5.3.166–71), ed. Kastan, 121.

33 Marlowe, *1 Tamburlaine the Great* (2.7.18-25), in *Complete Works*, ed. Fuller and Esche, 5:32

34 On the reframing of the Book of Nature as an impetus for scientific discovery in sixteenth-century Europe, see Peter Harrison, "The Book of Nature and Early Modern Science."

35 William Shakespeare, *Henry V* (2.4.104-5), in *The Norton Shakespeare*, ed. Stephen Greenblatt et al., 3d ed. All quotations from Shakespeare's works are to this edition, which uses the revised Oxford text, and will be cited hereafter parenthetically in the body of the chapter by act, scene, and line number. Quotes from *Hamlet* refer to the combined text of the second Quarto interpolated with the Folio.

36 For more on this clash, see Robert Miola's Introduction to his Norton edition of *Hamlet*, xviii.

37 Hazlitt, *Characters of Shakespeare's Plays*, 104.

38 Lee, *Shakespeare's "Hamlet" and the Controversies of the Self*, 158.

39 Greenblatt, *Renaissance Self-Fashioning from More to Shakespeare*, 87.
40 See Barker, *The Tremulous Private Body*, 31, 58, et passim.
41 On the importance of this passage for understanding the depth of Hamlet's subjectivity, see Barker, *The Tremulous Private Body,* 31, and Katharine Eisaman Maus, *Inwardness and Theater in the English Renaissance*, esp. 1–2.
42 Davies, *Imagining the Soul in Premodern Literature*, 163–72
43 Hundreds of medieval texts detailing "debate between the Soul and the Body" exist, including at least 132 copies of a manuscript entitled *Dialogus inter corpus et animam* sometimes attributed to Robert Grosseteste. For more on the context, see Robert W. Ackerman, "The Debate of the Body and Soul in Parochial Christianity." Michel-André Bossy mentions the ghost of Hamlet briefly in connection with this tradition in "Medieval Debates of Body and Soul," 45.
44 Greenblatt, "The Mousetrap," in *Practicing New Historicism*, 154.
45 Witmore, *Culture of Accidents*, 62. I would like to thank Jen Rust for bringing this study to my attention.
46 For more on this language, see Lewis, *Hamlet and the Vision of Darkness*, 43–111.
47 In the King James Bible, James 5:14-5 reads "Is any sick among you? Let him call for the elders of the church; and let them pray over him, anointing him with oil in the name of the Lord."
48 The sacrament was codified beginning in the eleventh century, especially in Origen and in the writings of Zachary of Besançon (d. 1155), who argued that "the visible anointing is the sacrament; but the reality of the sacrament is the remission which is conferred on the soul by the spiritual anointing of grace." See Boyd Taylor Coolman's chapter on "Twelfth Century Theology" in the *Oxford Handbook of Sacramental Theology*, 210. For an extended reading of this moment in relation to alchemy and questions about materialism, see my essay on "Some Versions of Enchanted Materialism: Bennett, Paracelsus, Hobbes, and *Hamlet*."
49 See Gad Freudenthal, "The Problem of Cohesion between Alchemy and Natural Philosophy," esp. 110–12, and his chapter on "Clandestine Stoic Concepts in Mechanical Philosophy" in *Renaissance and Reformation*.
50 See, e.g., the note to these lines in the *Norton Shakespeare*, ed. Greenblatt et al., 1820, n. 4. Some later editions of the play substitute "mattering" for "flattering," yet Q2, F and F2 texts of *Hamlet* all read "flattering."
51 Sherman, *Second Death*, 177.
52 I explore these associations between science, religion, and mimesis more fully in *Shakespeare's Theater of Nature*.

Part Two

Legacies of Early Modern Science and Theology

4

Versions of Negative Theology: Kenelm Digby and Margaret Cavendish

Peter Remien

Scholars have long struggled to reconcile Margaret Cavendish's orthodox religious beliefs with her daring and sometimes heretical natural philosophy.[1] In one of her *Philosophical Letters* (1664) devoted to differences between the "Natural Soul of Man," mutable and subject to corruption, and the "Divine Soul," Cavendish breaks off her speculations in sudden deferral to church authority: "for I should be loath to affirm anything contrary to their Doctrine, or the Information of the holy Scripture, as grounding my belief only upon the sacred Word of God and its true Interpretation made by the Orthodox Church."[2] Although Cavendish freely speculates about the physical properties of the Natural Soul—how it expands and contracts with the body, how it is physiologically diffuse and subject to deformities—once she turns to the immaterial Divine Soul, she abandons the spirit of philosophical inquiry. Indeed, this pivot from bold speculation to religious deference occasions a more general pronouncement about the limits of her natural philosophy. Addressing her fictional interlocutor, simply called "Madam," Cavendish writes, "I avoid, as much as ever I can, not to mix Divinity with Natural Philosophy; for I consider, that such a mixture would breed more confusion in the Church, then do any good to either; witness the doctrine of the Soul of Man, whereof are so many different opinions."[3] Cavendish's strict separation of natural philosophy from matters of religion, a gesture repeated throughout her philosophical oeuvre, would seem to fall in line with the broad and distortional narrative of the progressive secularization of science in early modern Europe that continues to structure popular conceptions of Baconian science. While Cavendish was a stylistic and intellectual virtuoso, whose playful and eclectic style enabled her capacious mind to move freely between disparate topics and literary genres, she was often hesitant to spell out the theological implications of her natural philosophy, particularly when they ran afoul of her professed orthodox Anglican beliefs.

And yet Cavendish's writing on natural philosophy was deeply theological, concerned with such foundational religious questions as the nature of free will, the origin of evil, and the relationship between creator and creation. It could hardly be otherwise. After all, seventeenth-century natural philosophy and religion were so intertwined that Amos Funkenstein uses the oxymoronic phrase "secular theologians" to refer to early scientific thinkers, meaning that they explored the divine structure of the universe outside the frameworks prescribed by religious institutions.[4] The pioneer of experimental science Robert Boyle has become a figurehead for the modern epistemological divide between the natural and social sciences.[5] But Boyle published a series of texts attempting to synthesize the new science with religious faith: *The Excellency of Theology Compared with Natural Philosophy* (1674), *Some Considerations about the Reconcileableness of Reason and Religion* (1675), and *The Christian Virtuoso: Showing That by Being Addicted to Experimental Philosophy, a Man Is Rather Assisted than Indisposed to Be a Good Christian* (1690). As these titles suggest, Boyle, like Cavendish, worried about potential conflict between his natural philosophical explorations and religious authority, but, unlike Cavendish, he actively sought to demonstrate their compatibility. Exiled and impoverished by the English Civil War and marginalized by her gender, Cavendish was more guarded in commingling science and religion. Lisa Sarasohn, in her important co-edited collection (with Brandie Siegfried) *God and Nature in the Thought of Margaret Cavendish*, highlights two entwined strategies that Cavendish uses to negotiate the theological underpinnings of her natural philosophy: fideism and negative theology.[6] Fideism, the claim that divine truths can be grasped only through faith, is embodied in the aforementioned passage from *Philosophical Letters*. Recognizing that her investigation of the soul has crossed into theological territory, Cavendish affirms her orthodox religious beliefs informed by scripture and adherence to church doctrine. But Cavendish's engagement with negative theology is broader and more encompassing. Negative theology—or apophatic theology, as it is also known, from the Greek for "denial" or "negation"—emphasizes the ineffability of God and the inability of human language, reason, and experience to capture the essence of divinity.[7] Present in different forms in Judaism, Christianity, Islam, and Buddhism, among other religious and philosophical traditions, negative theology is a "philosophical theology" encompassing metaphysics and ontology as well as religious practice.[8] Negative theology's formal development is often attributed to Pseudo-Dionysius the Areopagite, who fused Christian theology with Neoplatonic philosophy in the late fifth or early sixth century CE. It flourished in the Middle Ages in the Aristotelian theology of Thomas Aquinas and Anslem of Canterbury and later helped shape the modern existential philosophy of Friedrich Nietzsche, Søren Kierkegaard, and Martin Heidegger. Jacques Derrida and other postmodern philosophers draw heavily on negative theology in method if not in aims. As Catherine Keller describes, despite its apophatic interpretive framework, "deconstruction cannot be identified with negative theology, which remains, after all, a theology, indeed a theology indebted to the Neoplatonic One—of which poststructuralism is having *none*. Deconstruction is heir to the legacy of the death of God, the God of ontotheology whose Being *is* that One."[9] Ancient in origin, negative theology, with its apophatic approach to language and epistemology, remains a vital force in both sacred and secular forms of critical theory in the twenty-first century.

On the surface, it would seem that seventeenth-century natural philosophy would have little to do with negative theology—a practice rooted in mysticism and emotional experience rather than experimentation and rational inquiry. After all, the positivism of the new science clashes with the intellectual aporia of the apophatic. Francis Bacon's imperial project of "enlarging of the bounds of Human Empire, to the effecting of all things possible," expressed in his utopian fable *The New Atlantis* (1627)—a work parodied in Cavendish's *The Blazing World* (1666)—risks losing course in negative theology's cloud of unknowing.[10] What negative theology does offer early scientific thinkers like Cavendish—and her older contemporary Sir Kenelm Digby (whose work is explored in the next section)—is a way of mapping the threshold of divinity through the investigation of the material world. In his recent book *The Unknowable in Early Modern Thought*, Kevin Killeen notes the unexpected proliferation of "scientific-apophatic approaches" in the seventeenth century, positing that for both poets and natural philosophers, the "cusp of knowledge, what lay beyond the visible and beyond the sayable, was a valuable intellectual resource."[11] While Cavendish's vitalist material philosophy is rarely characterized as such, it is dualistic insofar as it envisions an immaterial God—as well as an immaterial Divine Soul—that exists beyond the processes of the material world. To be sure, Cavendish was deeply critical of René Descartes, one of the contemporary philosophers addressed in her *Philosophical Letters*, and her panpsychism—her theory that all matter is infused with the capacity for rational thought—self-consciously undermines the Cartesian divide separating mind from body and humanity from the rest of the material world. Cavendish was, however, consistent in her references to an ineffable and immaterial God whose ungraspable presence nevertheless proliferates in her natural philosophy. In Sara Mendelson's succinct formulation of Cavendish's negative theological natural philosophy, "Nature is known by her actions, while God is known for his absence."[12] Negative theology was therefore a necessary part of Cavendish's scientific thought—a way of exploring large cosmological questions about the nature and origin of the universe outside the strictures of church doctrine.

Cavendish's subtle but pervasive evocation of negative theology can also be understood through comparison with Digby. In the following section, I contrast the comprehensive natural philosophical system of Digby's *Two Treatises* (1644) with Cavendish's theory of rational self-moving matter, particularly as it is articulated in her mature philosophical works, *Observations upon Experimental Philosophy* (1666) and *Grounds of Natural Philosophy* (1668). While Digby's reductive mechanical philosophy supplies a material foil for the unique actions of the immaterial soul and a framework through which God acts upon the world, Cavendish's rational self-moving matter, which is infinite and eternal, mirrors her immaterial and ineffable God. In other words, Digby's negative theology is based on contrasting the material world with the immaterial realm of divinity, while Cavendish posits Nature as an imperfect but structurally adjacent surrogate for the God who eludes philosophical investigation. Indeed, as Cavendish's natural philosophy evolves, Nature increasingly eclipses the role of God and, in doing so, recapitulates God's ineffability. Nature, like God, cannot be understood by any single material being. With this in mind, the chapter's final section turns to the twenty-first century, pivoting from negative theology to what we might call "negative ecology"—modern ecology's tendency toward philosophical and material aporia. The idea is captured by Frank Egler's often quoted maxim, "ecosystems are not only more complex than we know, they are more complex than we can know."[13] This final section contrasts Digby's role in the development of ecology through his engagement with the seventeenth-century idea of "the oeconomy of nature"—a concept taken up by Carl Linnaeus and by both Erasmus and Charles Darwin—with Cavendish's less direct and influential, though perhaps philosophically deeper, ecological engagements. Cavendish's emphasis on the fragmentary, provisional nature of natural philosophical knowledge—the material mind's implication in Nature's autopoetic processes—presages the ecophilosophy of Timothy Morton as well as the new materialist political ecology of Jane Bennett.

The Science of Negative Theology

The natural philosophical systems of Digby and Cavendish offer distinctive visions of how the new science of the seventeenth century was used in service of negative theology. Members of the so-called Cavendish or Newcastle

Circle of natural philosophers in Paris (alongside Charles Cavendish, Walter Charleton, and Thomas Hobbes), Digby and Margaret Cavendish were among the first English writers to engage with the mechanical philosophy of Continental philosophers such as Descartes, Marin Mersenne, and Pierre Gassendi. Digby's opus *Two Treatises* was the first comprehensive work of mechanical philosophy published in English.[14] It was first published in Paris in 1644 and soon after in London in 1645 with subsequent editions in 1658, 1665, and 1669. The book's ambitious aim was theological in nature: to use the epistemological tools of the new science to prove the immortality of the soul. Encyclopedic in scope, the first treatise, *Declaring the Nature and Operations of Bodies*, is nevertheless reductive, attempting to explain all physical phenomena, including the seemingly purposeful actions of plants and animals, through simple mechanical principles. Digby's shorter second treatise, *Declaring the Nature and Operations of Man's Soul*, then proceeds in the negative, attributing to the immaterial soul observed phenomena that "cannot proceed from those [physical] principles" and therefore "cannot have a body for its source."[15] Digby's comprehensive system of mechanical philosophy thus serves the higher theological purpose of rationally and empirically demonstrating the limits of material explanation. The actions of the soul, Digby argues, cannot be attributed to material principles and therefore stem from an immaterial, and concordantly immortal, source. Linking the human with the divine, the soul marks the threshold of natural philosophy, the vanishing point at which scientific inquiry fails and religious awe takes over.

Digby begins the voluminous first treatise by explaining simple material phenomena such as a rock sinking in water before progressing to more complex cases like the seemingly rational actions of dogs and horses. Drawing largely upon Aristotle, though also engaged with the work of contemporaries like Descartes and Gassendi, Digby attempts to explain the collective actions of material bodies through the qualities of rarity and density, from which stem the principles of gravity and levity and the different properties of the four classical elements. Taken together "by a continued Progress, and not by leaps" like links in a great chain, as Digby describes in his commentary on Thomas Browne's influential *Religio Medici* (The Physician's Religion, 1643), Digby's observations demarcate a series of laws that govern the behavior of all material bodies.[16] Digby follows Descartes in understanding animals as mere "corporeal engines" (413) whose actions are determined by material causes—a point Cavendish refutes in many of her philosophical works. The theological trust of the first treatise is both to describe the intricate handiwork of "the Architect of nature" (317)—a metaphor to which we will return in the next section—and to articulate a set of principles that govern physical phenomena but fail to explain the unique nature of the soul.

Toward the beginning of the second treatise, Digby recapitulates the principles that he spent over four hundred pages outlining in the first treatise: "nothing whatsoever that we know to be a Body can be exempted from the declared Laws and orderly motions of Bodies" (353). It is worth noting that both Digby and Cavendish use the word "Body" (usually capitalized) in the specific seventeenth-century sense, common in alchemy and natural philosophy, of "substance" or "matter."[17] Such usage is commonplace in the period, but for Digby it has the effect of reinforcing the dualism of his argument, since the word most commonly refers to the human body, the material house and ontological counterpoint to the immaterial soul. Digby's obsession with the soul's immortality spanned his entire career as a natural philosopher. He first turned to experimental science after the death of his wife, Venetia, in 1633; his intense grief and preoccupation with the soul's immortality are eloquently documented in a manuscript copy of Digby's letters in Smith College's Mortimer Rare Book Collection MS 134. Digby's most comprehensive work of literary criticism, *Observations on the 22d Stanza in the 9th Canto of the 2d book of Spenser's "Faerie Queene,"* published in 1643, a year before the first edition of *Two Treatises*, interprets Spenser's geometric description of Alma's house as an elaborate allegory for the body's connection with the immaterial soul. While the soul's immortality may seem a fit subject for metaphysics and imaginative literature, the provenance of theological speculation rather than experimental investigation, Digby insists in his *Observations upon Religio Medici* that the soul's immortality is not simply "an Article of Faith" but "an evident Conclusion of demonstrative Science."[18] Initially published as a separate treatise in 1643, Digby's *Observations* were appended to the 1659, 1669, 1672, 1678, and 1682 editions of Browne's *Religio Medici*. In *Observations*, Digby praises Browne's philosophical speculations about the soul's immortality but criticizes his poetic conceits and eclectic style, urging instead a systematic exploration of the subject. This is, of course, precisely what Digby endeavors in his *Two Treatises*.

Digby begins his second treatise, like the first, with relatively simple examples that incrementally increase in complexity as the text unfolds. Cataloging the soul's abilities, Digby progresses from apprehension to understanding to knowledge to discourse to deliberative action and, finally, to rational thought, with each capacity building upon the last. He then explains why these capacities are unique to the soul and cannot proceed from the physical principles outlined in the first treatise. Digby contends that the soul differs from material bodies in its freedom from necessity, its role in ordering knowledge and things, and its capacity for infinite thought.

The soul's apprehension of infinity, Digby concludes, is particularly powerful evidence of its immortal nature: "Now if to this we add that what is to be known is absolutely and liquidly infinite, there cannot be brought or expected a more pregnant and home witness of our soul's spirituality". Unlike material bodies, the soul has "an infinite knowledge of infinite objects" (409). As the second treatise progresses, Digby assumes an increasingly mystical, apophatic tone, forgoing syllogisms and mathematical proofs for poetic reverie. Having "passed the Rubicon of experimental knowledge," Digby turns to metaphors and expressions of emotional ecstasy (415). Contrasting his observations with those of the first treatise, Digby writes, "I can find no possible means to link them together: a vast and impenetrable Ocean, lieth between the discoveries we have made on each side of its shores, which forbiddeth all commerce between them; at the least, on the dark bodies side, which hath not wings to soar into the region of *Intellectual* light" (414). While Digby's first treatise links together physical phenomena into an immense but rationally connected and thoroughly apprehensible chain—"the orderly causes, made and chained together" into a comprehensive whole—his second treatise denies the possibility of linking the soul to the physical world. The ontological gulf separating bodies from souls is "a vast and impenetrable Ocean" that cannot be crossed by physical bodies (414).

Digby concludes his second treatise with an extended apostrophe to his immortal soul, personified as "her," who becomes the object of ecstatic praise. Lost in apophatic contemplation, Digby attempts to express the inexpressible through emotionally charged poetic language:

> How is it possible, then, for my narrow heart to frame an apprehension of the infinite excess of thy pleasure (my soul) over all the pleasure this limited world can afford, which is all measured by such petty proportions? How should I stamp a figure of thy immense greatness into my material imagination? Here I lose my power of speaking, because I have too much to speak of: I must become silent and dumb because all the words and language I can use express not the thousandeth nor the millionth part of what I evidently see to be true. All I can say is that, whatsoever I think or imagine, it is not that: and that it is not like any of those things; unto some of which unless it be like, it is impossible for me to make any proportion or similitude unto it. What then shall I do but lay myself down in mine own shadow and there rejoice that thou art a light so great as I am not able to endure the dazzling splendor of thy rays: that thy pleasure is so excessive as no part of it can enter into my circumscribed heart without dilating it so wide that it must break in sunder: and that thy happiness is so infinite as the highest pitch I

> can hope for to glut myself withal during this dark night of my tedious pilgrimage here on earth is to see, evidently, that it is impossible for me in this life to frame any scantling of it; much less, to know how great it is. (454–5)

At this point in the text, Digby the eminently rational natural philosopher has transformed himself into a Christian mystic who can only marvel at the soul's immense beauty and infinitude. Imagining the moment of his death as a joyous revelation, Digby gestures toward the limits of human knowledge. In the first treatise, Digby had made extensive use of algebra and geometry, but in the second he evokes the limits of mathematical knowledge. While Digby's forty years of life had been filled with great suffering—his father, Everard Digby, was executed in 1606 as part of the failed November 5 Gunpowder Plot; he lived a life of persecution and exile because of his Roman Catholic faith; and his wife, Venetia, died under mysterious circumstances at the age of 33—his soul will eventually experience pleasure "multiplied beyond the Arithmetic of intelligences" (463).[19] Grappling with the implications of the soul's immortality "which surpasseth all multiplication," Digby writes, "I see plainly that I must multiply eternity by eternity, to frame a scantling of that bliss" (463). In the conclusion's final paragraph, Digby turns not to Aristotle, his intellectual lodestar, but to "the mystagogical illuminations of the great Areopagite"—the Pseudo-Dionysus widely understood to be the founder of the Christian apophatic tradition. The conclusion's final sentence is an act of undoing, the negation of Digby's life work, his entire 464-page tome; in a self-consciously futile gesture, it attempts to imagine the soul's "true bliss, in comparison whereof, all that we have hitherto declared is but shadow, vanity, and nothing" (464).

The negative theology of Cavendish is simultaneously more subtle and more pervasive than that of Digby. She makes no grand pronouncements about the theological ambitions of her natural philosophy, but her philosophical writings abound with references to a God who simultaneously transcends and parallels Nature—her quasi-deistical name for collective material phenomena. Digby was friends with Cavendish's husband, William Cavendish, First Duke of Newcastle, and part of the aforementioned intellectual circle centered on the Cavendish household (known alternatively as the "Cavendish Circle" or the "Newcastle Circle") comprising Continental philosophers such as Descartes and Gassendi, together with Royalist exiles from England like Digby, Hobbes, and Walter Charleton. My point in juxtaposing the natural philosophy of Digby and Cavendish is less to suggest a direct line of filiation between them—though they were certainly familiar with each other's writings—than to illustrate two distinct ways that negative theology can function in seventeenth-century natural philosophy, one based on distinction and the other on adjacency. A founding member of the Royal Society, an institution from which Cavendish was excluded on the basis of her sex, Digby, however eccentric he appears to twenty-first-century readers, might be taken as paradigmatic of seventeenth-century natural philosophy with its theological preoccupations and intense rationalization (and mechanization) of the natural world. Digby synthesized the new science of Bacon, Descartes, and Boyle with the old Aristotelian physics while maintaining a resolutely theological focus. Although intended as a unique honor, Cavendish's 1667 visit to the Royal Society, hosted by Boyle, Robert Hooke, and others, served to reinforce her exclusion from the bastions of male intellectual society. Indeed, much of Cavendish's natural philosophy, with its criticism of the period's fashionable scientific ideas and instruments, is articulated against that of her more privileged male counterparts.

Digby, though, seemed to have admired Cavendish's philosophy. In a letter he wrote to Cavendish from Paris dated June 9, 1657, he praises her as "the Ornament of this" age and suggests that she "will be the Envy of all future ones." Focusing on the content of her book, Digby continues, "every Page in your excellent Book affordeth abundant matter."[20] It is difficult to say which book Digby is referencing. At this point in her career, Cavendish had already published five books, including *Philosophical and Physical Opinions* (1655), in which she articulates, in nascent form, her theory of rational self-moving matter. While Digby's allusion to the book's "abundant matter" is, no doubt, simply a reference to the work's content, whatever that content may be, it is tempting, given Cavendish's career-spanning obsession with the nature of the material world, to interpret it as a reference to her bold and original theory of matter.

Under the influence of the mechanisms of Descartes and Gassendi and the revival of Epicurean atomism following the fifteenth-century rediscovery of Lucretius's *On the Nature of Things*, seventeenth-century natural philosophers like Digby understood matter as inert bodies moved by external forces. Cavendish, however, suggests that matter is not only animate and self-directed but also capable of rational deliberation. Positing a "Triumvirate of the degrees of Matter," Cavendish divides matter into rational, sensitive, and inanimate parts, each of which is responsible for a different function.[21] These three "degrees," as Cavendish consistently refers to them, constitute differences in degree rather than type, since Cavendish maintains that all matter is unified in a single body, which she refers to as Nature. Like the mysterious Holy Trinity at the center of Christianity, Cavendish's matter is somehow both tripartite and one, and thus Cavendish's philosophy is often described as "monistic vitalism," despite matter's internal divisions.[22] Toward the beginning of *Observations upon Experimental Philosophy*, Cavendish uses the example of a rider on a horse carrying a stick to illustrate her theory. The horse represents matter's sensitive degree, the rider matter's rational degree, and the stick the inanimate degree. While it is the rider who guides the horse, which creates the motion, the stick too moves with the other objects. Orchestrated together, the three degrees of matter constitute a unity that ensures Nature's orderly functioning (25-6). (Cavendish's other preferred metaphor for matter's three degrees—the architect, the worker, and the material—will be explored in the next section.) The radical implication of Cavendish's theory of matter is Nature's autonomy, her creative ability to construct and order the universe on her own, without an external force. Digby had envisioned the ability to instill order as a unique capacity of the immaterial soul, but Cavendish represents matter as capable not only of internal motion but also of self-organization.

But where and how does God fit into Cavendish's natural philosophy? What is God's role in the functioning of the material world? How, in other words, does God relate to Cavendish's quasi-deistical Nature? After all, matter's capacity for motion, self-organization, and rational deliberation would seem to render redundant any external deity. Cavendish is equivocal in this regard. She continuously refers to Nature as God's "eternal servant" and emphasizes her inferiority when compared with his perfection, and yet God has the tendency to recede into the background of Cavendish's natural philosophy. Sarasohn explains how, as Cavendish's materialist philosophy matured, it "seemed to reduce the action of God to the noninterventionist passivity of the Epicurean gods."[23] Since God is immaterial and ineffable, beyond the reach of human knowledge, he can only be explored in the negative. Cavendish, like Digby, tends toward dialectic expositions of divinity. The dualism of *Two Treatises* is formally instantiated in its separation into two books, one devoted to material bodies, the other to the immortal soul. Cavendish's dialectics of God and Nature tend to be woven more intricately into the fabric of her texts. *Observations upon Experimental Philosophy*, for example, begins with a dialogue between the author's "former thoughts" and her "latter thoughts," the latter of which answers questions posed by the former. *Philosophical Letters*, likewise, unfolds as a fictional epistolary dialogue between an unnamed "Madam," who wishes to know more about the author's natural philosophy, and a fictionalized version of Cavendish the author. Such dialogues embody Cavendish's idea of knowledge as always provisional and in process, and they provide appropriate frameworks for Cavendish's negative theology. A short chapter in Cavendish's most mature philosophical work, *Grounds of Natural Philosophy*, titled "Of the Differences between God, and Nature," is paradigmatic in its dialectical approach to the relationship between God and Nature:

> GOD is an Eternal Creator; Nature, his Eternal Creature. GOD, an Eternal Master: Nature, GOD's Eternal Servant. GOD is an Infinite and Eternal Immaterial Being: Nature, an Infinite Corporeal Being. GOD is Immovable, and Immutable: Nature, Moving, and Mutable. GOD is Eternal, Indivisible, and of an Incompoundable Being: Nature, Eternally Divisible and Compoundable. GOD, Eternally Perfect: Nature, Eternally Imperfect. GOD, Eternally Inalterable: Nature Eternally Alterable. GOD, without Error: Nature, full of Irregularities. GOD knows exactly, or perfectly, Nature: Nature doth not perfectly know GOD. GOD is Infinitely and Eternally worshipped: Nature is the Eternal and Infinite Worshipper.[24]

Although God appears as Nature's opposite in this lengthy and parataxic list—the perfect creator of an imperfect creation—the repetition of the words "Eternal" and "Infinite" foreground important similarities between them.[25] Cavendish's Nature is, like God, infinite and eternal. For Digby, the infinite and immortal nature of the soul marked the threshold of human knowledge and the dividing line between the material world of nature and the immaterial world of divinity. For Cavendish, though, Nature and God are less opposites than coextensive parts of a whole, each extending infinitely in time and space. The infinite and the eternal are, of course, unknowable, and this poses a problem not only for Cavendish's theology but also her natural philosophy.

Cavendish addresses God's unknowability in several of her philosophical works. In *Observations upon Experimental Philosophy*, for example, she asserts that it is impossible for any "particular natural Creatures" to "have an exterior perception of the substance of God." Because of his immaterial nature, God lies beyond the purview of perception and therefore outside the realm of natural philosophy. What natural creatures do possess, however, is "an interior self-knowledge of the existency of the Eternal and Omnipotent God, as the Author of Nature." Reiterating this point as part of the aforementioned dialogue between her "former thoughts" and "latter thoughts," Cavendish continues, "the knowledge of the existency of God, as I mentioned before, is innate and inherent in Nature and all her parts as much as self-knowledge is" (16-17). The human capacity to comprehend divinity is, therefore, limited, a matter of intuitive self-understanding rather than empirical observation. Cavendish's separation of the imperceivable "substance of God" from the intuited knowledge of God's "Existency" exemplifies the negative theological "distinction between knowing 'that' God is and knowing 'what' God is."[26] Natural philosophy thus leads to apophatic resolution about a deity whose existence is inherently understood but whose nature eludes investigation.

The more radical implication of Cavendish's natural philosophy is the possibility of the nonhuman investigation of divinity. Indeed, Cavendish's term "particular natural Creatures" is expansive, encompassing not only humans, but also animals, plants, minerals, and all other material beings. Scholars describe this philosophy as panpsychism—the dispersal of rationality throughout the natural world. Cavendish's natural philosophy, her attempts to understand Nature and the God of Nature, tends to be more empathetic, imaginative, and collaborative than Digby's vast mechanical system. Cavendish returns to the inherent limits of the observations of "particular" creatures throughout her writings. Midway through the first part of *The Blazing World*, the Empress, rejecting the distortional effects of human artifice, concludes that "Nature's Works are so various and wonderful that no particular creature is able to trace her ways."[27] Here, the formula that Cavendish had previously applied to God in *Observations upon Experimental Philosophy* pertains also to Nature. If the infinite God is ineffable, then so too is infinite Nature, at least from any particular vantage point. Cavendish implements this notion of the creaturely fragmentation of knowledge in the *Blazing World*'s proliferation of human-animal hybrids—Bear-men, Fox-men, Bird-men, Lice-men, Fish-men, Worm-men, Ape-men, and the like—who collectively (and imperfectly) explore the workings of Nature. These hybrid creatures allegorize different kinds of contemporary natural philosophers, but they also represent more generally the creaturely diversity of perspectives—the epistemological fragmentation that renders impossible any comprehensive understanding of Nature.

Negative Ecology

The final section of this chapter turns from the exploration of negative theology to what I call "negative ecology"—the philosophical and practical limits of ecological thought.[28] In acknowledging the complexity and diversity of Nature, Cavendish prefigures the roles of chaos theory and disequilibrium in twenty-first-century ecology, as well as the influential ecophilosophy of Timothy Morton.[29] Like the unrepresentable vision of evolution based on horizontal gene transfer that is emerging in the twenty-first century, Cavendish's natural philosophy posits an unknowable version of Nature characterized by infinite variety and unpredictable autopoetic processes. As David Quammen describes in *The Tangled Tree*, horizontal gene transfer, the non-sexual passage of genes between bacteria and other forms of life, upends the older evolutionary metaphor of a tree of life.[30] In the context of this chapter, negative ecology is also meant to suggest the theological dimensions of ecological thought, which are deeply engrained in the history of ecology. To this point, I contrast Digby's role in the development of ecology—through his establishment of "the oeconomy of nature"—with Cavendish's proto-ecological understanding of Nature.

Given Cavendish's empathetic imagination, her opposition to vivisection, her rejection of Cartesian dualism, and her serious attention to nonhuman perspectives, it is unsurprising that modern scholars have come to appreciate her contributions to ecological thought. In her pioneering ecofeminist study, *Speaking for Nature*, Sylvia Bowerbank foregrounds the ecological dimensions of Cavendish's empathetic natural philosophy as well as her focus on environmental stewardship in the wake of the English Civil War.[31] Sylvia Bowerbank argues that the conventional association of women with nature and their simultaneous inclusion and exclusion from the category of "man" afford early modern women like Cavendish the ability to speak on nature's behalf. To this point, Diane Kelsey McColley highlights how Cavendish's attention to nonhuman voices in poems such as "The Hunting of the Hare" and "The Earth's Complaint" enable her to oppose ecologically damaging practices such as aristocratic hunting parties.[32] More recently, Elizabeth Gruber has positioned Cavendish as the foremost embodiment of the "eco-self"—an idea that "conveys the relational dimensions of humanness, our perpetual [...] embeddedness in the surrounding world" while acknowledging "the rich store of aesthetic and cultural endeavors that define humans" and embracing the "concordant need to claim a unique vantage point."[33] In *The Blazing World*, Gruber argues, Cavendish "mates a fierce individualism with the realization that person and world are indissolubly linked."[34] As Gruber's study illuminates, Cavendish's materialist philosophy, with its material theory of mind and dispersal of rationality throughout Nature, troubles the self/world distinction.

The history of ecology has always been deeply entrenched in theological ideas. While the science of ecology was not formally established until the late nineteenth century—and didn't mature as a discipline until the mid-twentieth—the protoecological idea of "the oeconomy of nature" was first developed in the mid-seventeenth century by Digby.[35] For Digby, the oeconomy of nature is a thoroughly theological concept that describes the divinely structured arrangement of the material world. In the early modern sense, oeconomy—from the Greek parts *oikos* (house) and *nomos* (law)—did not mean political economy—as in our modern sense of *the* economy—but rather household management.[36] Digby first mentions the oeconomy of nature in *Two Treatises* in his reconceptualization of Aristotelian natural motion. He describes how "bodies that of themselves have no propension unto any determinate place do nevertheless move constantly and perpetually one way, the dense ones descending and the rare ones ascending: not by any intrinsic quality that worketh upon them, but by the oeconomy of nature that hath set on foot due and plain causes to produce known effects" (83). While Aristotle understood material bodies as containing within themselves the principles of motion, Digby envisions an external force—the oeconomy of nature—responsible for their collective movements. This force was, of course, divine in origin, a system set into motion by God's provident architecture. Notably, Digby's system is entirely predictable—bodies move "constantly and perpetually one way"—and thoroughly knowable. As we have seen, the predictable and knowable world of material bodies provides Digby with a foil for the immaterial actions of the immortal soul. Digby's oeconomy of nature was thus the work of "the Almighty Architect" whose "handiwork [can be seen] in every particular effect of nature" (227). As Donald Worster describes in his influential history of ecology, "God's economy was His extraordinary talent for matching means to ends, for so managing the cosmos that each constituent part performed its work with stunning efficiency."[37] While ecology's foundational figure, Charles Darwin, has become a figurehead for secularization, the economy of nature (a phrase Darwin used in lieu of ecology to describe systemic interactions in nature) retained its theological origins through much of the nineteenth century. The seeds of its secularization, though, may already have been planted. As Peter Harrison has recently argued, "the Protestant reformers' attempt to make the theological domain all-encompassing ultimately gives rise to a naturalistic re-description of that theological totality."[38] Digby was a Catholic, but Harrison's description of the ironic process of the universal sacralization of nature followed by secularization is an apt description of the theological oeconomy of nature's later development into the science of ecology.

Like Digby, Cavendish uses architectural metaphors in her description of Nature, but they lead to a different portrait of the material world. Illustrating the previously described rational, sensitive, and inanimate degrees of matter, Cavendish turns to Aristotle's (and Digby's) favorite metaphor: the construction of the house. In *Observations Upon Experimental Philosophy*, she explains how

> those three degrees were so inseparably commixt in the body of Nature that none could be without the other in any part or Creature of Nature, [even if it] could [...] be divided to an Atom. For, as in the Exstruction [i.e., building] of a house, there is first required an Architect or Surveyor who orders and designs the building and puts the Laborers to work, next the Laborers or Workmen themselves, and lastly the Materials of which the House is built: so the Rational part, said they, in the framing of Natural Effects, is, as it were, the Surveyor or Architect; the Sensitive, the laboring or working part; and the Inanimate, the materials; and all these degrees are necessarily required in every composed action of Nature.
>
> (24)

Although Digby and Cavendish describe the material world in similar ways, employing the kind of metaphor that the philosopher Michael Zimmerman refers to as "productionist metaphysics," they reach nearly opposite conclusions about the relationship between God and Nature.[39] For Digby, God's architecture provides the framework for nature's orderly and predictable functioning. For Cavendish, however, Nature is her own architect—as well as her own worker and materials. Her autopoetic processes function without the predictability of Digby's system. In Cavendish's account, the human mind, the ostensible locus of rationality, is itself implicated in Nature's processes. Indeed, as Liza Blake points out in her incisive essay on Cavendish's relationship with the new materialism—particularly the vibrant matter of Jane Bennett—"Cavendish uses her vitalist materialism to argue posthumanist points, but more importantly she does so by showing that any vitalist posthumanist project will fail as long as it relies on a definition of life derived from, or synonymous with, human life or human agency."[40] Refusing to privilege human agency, or even life, Cavendish offers a radically decentered view of an unpredictable and unknowable material world that is unconstrained by God or human epistemological mastery. Nature's house, of which we are all part, remains a mystery.

Cavendish's emphasis on the limits of human knowledge not only of divinity but also of Nature itself prefigures the ecophilosophy of Timothy Morton, whose Derridean approach foregrounds the complex theoretical implications of the science of ecology.[41] Morton's philosophy, like Cavendish's, abounds in paradoxes. To think ecologically, Morton suggests, is to acknowledge that "there is no metaposition from which we can make ecological pronouncements."[42] Indeed, throughout their oeuvre Morton rejects many of the distinctions that make objective thought possible: nature/culture, subject/object, inside/outside, self/other. We are instead enmeshed in a world that baffles comprehension because of our involvement in it. This is also, of course, the world of Cavendish's natural philosophy, where (as we have seen) an ineffable God gives way to an ineffable Nature, whose infinite material body encompasses our very thoughts. In many ways, the culmination of Morton's theory lies in their notion of hyperobjects: "things that are massively distributed in time and space relative to humans."[43] Hyperobjects include things like global warming, ecosystems, and nuclear waste—objects that are objectively real and yet cannot be objectively observed because of their inhuman temporal and spatial scope. Derived in part from the Object-Oriented Ontology of Graham Harman and others, where objects exist in a state of "withdrawal," hyperobjects represent the eerie unknowability of the Anthropocene, the inconceivable destructiveness of the present historical moment. Although Morton argues that rejecting the concept of "Nature" as humanity's metaphysical other is a necessary precursor to properly ecological thought, Cavendish's idea of Nature as the infinite and eternal totality of the material universe is an apt embodiment of Morton's idea of the incomprehensible but materially present hyperobject. If Digby represents one lineage of ecology based on the assumption of a more or less predictable and knowable nature—that which can be quantified and empirically observed—Cavendish prefigures ecology's radical unmooring of human knowledge evident in the work of thinkers like Morton—a negative ecology based on the world's ineffable complexity.

Notes

1 See, for example, Hilda L. Smith, "Claims to Orthodoxy," 15–26.
2 Cavendish, *Philosophical Letters*, 210. Unless otherwise noted, all references to Cavendish are original publications accessed through Early English Books Online. Though it is not available at the time of this chapter's composition, Liza Blake, Shawn Moore, and Jacob Tootalian are currently working on the twenty-volume *Complete Works of Margaret Cavendish*, a collection of critical editions under contract with Punctum Books: http://digitalcavendish.org/complete-works/.
3 Cavendish, *Philosophical Letters*, 210.
4 Funkenstein, *Theology and the Scientific Imagination*, 3.
5 See Steven Shapin and Simon Schaffer, *Leviathan and the Air-Pump;* and Bruno Latour, *We Have Never Been Modern*.
6 Lisa T. Sarasohn, "Fideism, Negative Theology, and Christianity in the Thought of Margaret Cavendish," 93–106.
7 For a good overview of the history of negative theology see Ilse Bulhof and Laurens ten Kate, *Flight of the Gods*.

8 Ibid., 4.
9 Keller, *Cloud of the Impossible*, 8, emphasis in original. For Jacques Derrida's engagement with negative theology, see *Derrida and Negative Theology*, ed. Coward and Foshay.
10 Bacon, *New Atlantis*, cited in Cavendish, *Paper Bodies*, ed. Bowerbank and Mendelson, 264.
11 Killeen, *The Unknowable in Early Modern Thought*, 2.
12 Sara Mendelson, "The God of Nature and the Nature of God," 31.
13 Frank Egler, quoted in *Ecosystem Management*, ed. Boyce and Haney, 2.
14 For background on the "Cavendish" or "Newcastle Circle" see Robert H. Kargon, *Atomism in England from Hariot to Newton;* and Steven Clucas, "The Atomism of the Cavendish Circle: A Reappraisal."
15 Digby, *Two Treatises,* sig. ú1. The original edition of 1644 is hereafter cited parenthetically by page number in the main text.
16 Digby, *Observations upon "Religio Medici,"* 15.
17 *Oxford English Dictionary*, "body (n.)," def. 5.
18 Digby, *Observations upon "Religio Medici,"* 12.
19 For good biographies of Digby see Robert Peterson, *Sir Kenelm Digby*; and Joe Moshenska, *A Stain in the Blood.*
20 Digby et al. *A Collection of Letters and Poems*, 65.
21 Cavendish, *Observations upon Experimental Philosophy*, ed. O'Neill, 25. All references to this work are to this edition, hereafter cited parenthetically in the text by page number.
22 See, for example, John Rogers, *The Matter of Revolution*, chapter 6.
23 Sarasohn, *The Natural Philosophy of Margaret Cavendish*, 67.
24 Cavendish, *Grounds of Natural Philosophy*, 241.
25 In "Margaret Cavendish and the Nature of Infinity," Mendelson positions Cavendish's career-spanning fascination with infinity alongside seventeenth-century scientific developments like the discovery of Calculus, ultimately arguing that thinking about infinity expanded Cavendish's natural philosophy into a "theory of everything" (in *God and Nature in the Thought of Margaret Cavendish*, ed. Siegfried and Sarasohn, 77).
26 Bulhof and ten Kate, *Flight of the Gods*, 25.
27 Cavendish, *The Blazing World*, in *Paper Bodies*, 185.
28 The term "negative ecology" is used in different contexts in a couple of recent studies. In David Bond's recent book, *Negative Ecologies,* negative ecology refers to how many of the recent scientific discoveries about the environment have been facilitated by fossil fuel culture, which contributes to destroying the ecologies that it renders visible. In Seth McGrath's book *Thinking Nature: An Essay in Negative Ecology*, the phrase refers to the philosophical movements that emerge in the wake of nature's conceptual and material destruction. By negative ecology, I simply mean the epistemological limits of ecological inquiry, while suggesting structural parallels with the philosophy of negative theology.

29 For an overview of post-equilibrium ecology, see Daniel Botkin, *Discordant Harmonies.*
30 See David Quammen, *The Tangled Tree: A Radical New History of Life.*
31 Bowerbank, *Speaking for Nature.*
32 McColley, *Poetry and Ecology in the Age of Milton and Marvell.*
33 Gruber, *The Eco-Self in Early Modern English Literature*, 16.
34 Ibid., 179.
35 See Worster, *Nature's Economy: A History of Ecological Ideas;* and Remien, *The Concept of Nature in Early Modern English Literature.*
36 For oikonomia's role in the development of modern political economy, see Maifreda, *From Oikonomia to Political Economy*; Margaret Schabas and Neil de Marchi *Oeconomies in the Age of Newton*; and Schabas, *The Natural Origins of Economics.*
37 Worster, *Nature's Economy*, 37.
38 Harrison, "Science and Secularization," 58.
39 Zimmerman, *Heidegger's Confrontation with Modernity*, 157.
40 Blake, "After Life in Margaret Cavendish's Vitalist Posthumanism," 436.
41 Over the past decade and a half, Timothy Morton has developed and refined their ecophilosophy in a number of important books including *Ecology Without Nature*, *The Ecological Thought*, *Hyperobjects*, *Dark Ecology*, and, most recently, *Hell: In Search of a Christian Ecology.*
42 Morton, *The Ecological Thought*, 17.
43 Morton, *Hyperobjects*, 1.

5

The Religious Origins of Colonial Science: Hans Sloane as Bacon's New Solomon

Amy Cooper

Introduction

In an entry titled "Mushrooms, Mosses, etc." in his *Voyage to Jamaica* (1707), Sir Hans Sloane glosses a passage from 1 Kings about King Solomon: "this kind of plant growing on Walls is with some reason supposed to be that which is translated Hyssop in our Bibles, where Solomon is said to have spoken of trees, from the Cedar of Lebanon, to the Hyssop on the Wall."[1] Sloane's reference to Solomon belongs to a natural-historical tradition that reaches back through Francis Bacon in England and Marsilio Ficino in Italy to the *Hermetica* of ancient Egypt. The reference to the hyssop in 1 Kings 4:33 had long served as evidence for the authority and importance of natural history. In the early modern period, Solomon's knowledge of trees "from the cedar [...] to the hyssop" was interpreted by natural philosophers such as Bacon to mean that King Solomon was himself a natural historian and had amassed a lost body of botanical knowledge.[2] For this reason, Bacon invoked Solomon as a symbol of his *Great Instauration* (1620), which Bacon grounded in a new and comprehensive natural history. In his dedicatory letter, Bacon fashions King James as a new Solomon who might "rival that same king by putting in hand the collecting and perfecting of a true and rigorous natural and experimental history which (stripped of philological matters) may be designed for the building up of philosophy" (*Organum*, 9). When he names the state-funded research foundation of the *New Atlantis* (1627) the "House of Salomon," Bacon has both the biblical passage and James himself in mind: the House of Solomon distills Bacon's vision for a renewal of natural philosophy, which he hoped would be managed by a state-sponsored organization devoted to collecting the natural wonders of the world. Such a vision would not come to pass until Charles II granted the Royal Society a charter in 1662—and even then, the Royal Society was never a true fulfillment of Bacon's House of Solomon.[3] But its fellows frequently invoked Bacon, in whose image they founded the Society itself and whose championing of natural history authorized projects like Sloane's *Voyage*.

As Lord Chancellor under King James, Bacon had extensive ties to colonizing projects, including the Virginia Company, Newfoundland Company, and East India Company. Bacon even had a stained-glass representation of a tobacco plant and turkey made for his Gorhambury estate.[4] These biographical details suggest affinities with Sloane, who owned and operated a Jamaican plantation that relied on slave labor to produce sugar. Bacon's colonial interests extended also to his philosophical and scientific work; indeed, the metaphorical architecture of Bacon's writings invoke the tools and technologies of colonial empire.[5] The *Novum Organum* or "new instrument" with which Bacon hopes to sail away from the familiar coasts of ancient knowledge and into the unexplored oceans of the new philosophy is a navigational compass—an instrument that enabled colonial expansion. The frontispiece of *The Great Instauration* features a warship sailing through the Pillars of Hercules, symbols of the limits of the known world (see Fig. 5.1). It links navigation and exploration of the new world with the project of natural philosophy through an inscription from the Book of Daniel, "*Multi pertransibunt et augebitur scientia*" ("many shall pass through, and knowledge shall be increased").[6] Sloane, seeking to draw upon Bacon's authority in establishing his own as a natural historian, also adopted this verse from Daniel in the title page to the *Voyage*.[7]

Although scholars working in American, Caribbean, and circum-Atlantic literatures acknowledge Sloane's self-styled Baconianism, he is often read forward into subsequent histories of science. Many of his specimens are still housed at the British Natural History Museum and he helped to found the British Museum.[8] Bacon has occupied a similar position in histories of science that tend to see him as the origin point in forward-looking histories of not just experimental science but also of modern and ostensibly secular science.[9] Reading Sloane backward through Bacon to the Hermeticism of classical antiquity, understood less as a magical than as a religious tradition, instead reveals a different genealogy of both capitalist extraction and colonial exploitation, revealing how both enterprises were embedded in the foundational narratives of the new science from the start. Ficino's translation of newly recovered Hermetic texts in the fifteenth century, erroneously believed to be contemporary with the Mosaic texts of the Bible, underlies Bacon's belief that natural philosophy would produce the promised restoration of the fallen, Hermetic Adam.[10] In fashioning himself as a Baconian natural historian, Sloane places himself in an intellectual genealogy that is Hermetic in origin.

Figure 5.1 Simon de Passe, frontispiece to Francis Bacon, *The Great Instauration*, 1620. © The Trustees of the British Museum. Shared under a Creative Commons Attribution-NonCommercial-ShareAlike 4.0 International (CC BY-NC-SA 4.0) license.

The first part of this essay traces the influence of religious Hermeticism on Bacon's theologically inflected vision of natural history. The second half connects Sloane to this minor religious tradition by offering a new reading of the specimen plates that divide volume one of the *Voyage*. Building on the work of Kay D. Kriz, Christopher Iannini, and James Delbourgo, I argue that the "Carvell" or Man-of-War jellyfish that appears in one of Sloane's specimen plates next to coral-encrusted coins and a piece of wood from a sunken Spanish galleon (see Fig. 5.2) functions as an intertextual reference to the emblems in the frontispiece of Bacon's *Great Instauration*.

Specifically, Sloane's Carvell shares formal resemblances with the ship that appears on Bacon's frontispiece. As an emblematic device, moreover, Sloane's specimen plate consolidates the Caribbean plantation as the site through which Bacon's new science will be realized by its eighteenth-century practitioners. Thus, the intertextual links between Bacon's emblematic frontispiece and Sloane's emblematic engraving showcase the specimen as the

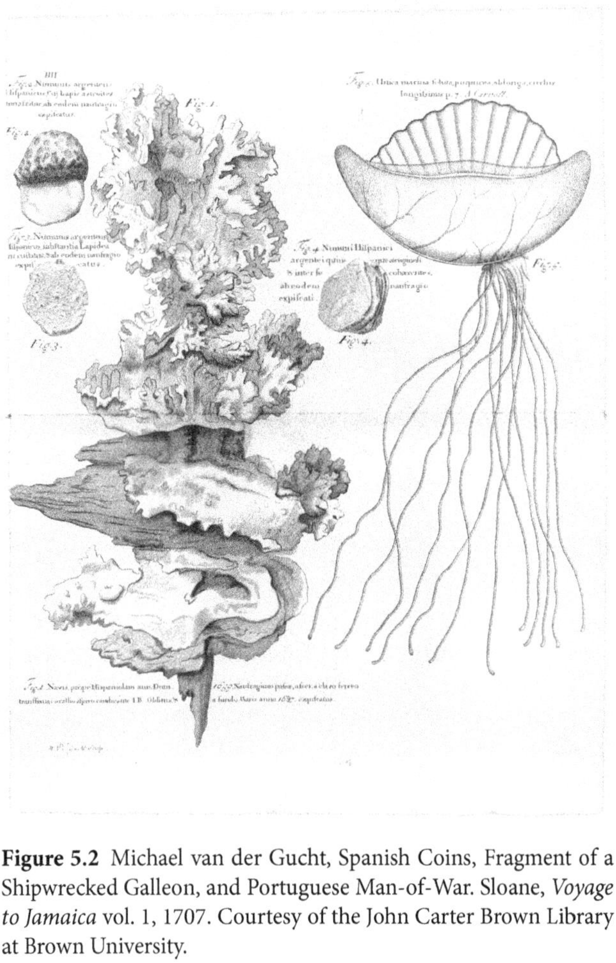

Figure 5.2 Michael van der Gucht, Spanish Coins, Fragment of a Shipwrecked Galleon, and Portuguese Man-of-War. Sloane, *Voyage to Jamaica* vol. 1, 1707. Courtesy of the John Carter Brown Library at Brown University.

device or instrument—a different kind of *novum organum*—through which the objects and information flowing from the Caribbean can be reconstituted as scientific facts. These specimen-emblems call upon the foundational narratives of Bacon's new science, which themselves rely on the authority of the Solomon myth, as part of the authorizing function of the scientific fact described by historians of science such as Delbourgo and Steven Shapin.[11] This essay argues for the centrality of Hermeticism and the Solomon myth to the practice of natural history after Bacon and traces the ways in which this alternative religious history came to invisibly underwrite the brutal system of slavery and colonial exchange fueling the new science.

Francis Bacon and the Hermetic Adam

Recent studies of Bacon and religion have rightly taken the Protestant Reformation as the most salient context for understanding the religious dimensions of his natural philosophy. Kristen Poole has connected developments in religious studies, nuancing our understanding of Protestant allegory to challenge narratives about Bacon's scientific secularism.[12] Travis DeCook has traced the imagery of arks in Bacon's writings to explain the strange, Pentecostal conversion narrative of the Bensalemites in the *New Atlantis*.[13] Much less attention has been paid to the influence of Hermeticism on Bacon's understanding of scripture, yet Bacon's conception of religious history was as steeped in magic traditions as his science. In the last century, Frances Yates, D. P. Walker, and Paolo Rossi brought Hermeticism to the attention of scholars in an effort to recover magic traditions—alchemy, astrology, cabala, Neoplatonism, etc.—and to challenge the exclusion of magic from the history of science.[14] In what later became known as the Yates Thesis, Yates argued that the Hermetic magus of the sixteenth century prefigured the Christian virtuoso of the seventeenth century and that the experimental program of the seventeenth century had its roots in the operative powers promised by Hermetic philosophy.[15] While Giordano Bruno served as Yates's exemplary figure of the Hermetic magus, Rossi pointed to Bacon as an important liminal figure in the transition from the Hermetic philosophy of the sixteenth century to the new experimental science of the seventeenth.[16] This section argues that two distinct religious logics inform Bacon's body of work: one Judeo-Christian and the other Hermetic, although, like his contemporaries, Bacon saw them as part of a single, unified tradition—what D. P. Walker calls the *prisca theologia* or "ancient theology."[17] In a gloss citing Walker on Bacon's association of Solomon with natural history in the *Advancement of Learning* (1605), Michael Kiernan describes "the doctrine of *prisca sapientia*" or "ancient wisdom" as a common belief among Hermeticists that "revelation could be parsed in the writings of certain inspired ancients (e.g., Hermes Trismegistus, Orpheus)" (*Learning* 235).[18]

Bacon repeatedly labors to connect these two traditions in his *Advancement of Learning.* As Bacon ranges through various biblical supports for the pursuit of natural philosophy, he contends that Moses "is adorned by the Scriptures with this addition, and commendation: That he was seen in all the learning of the Egyptians, which nation we know was one of the most ancient schools of the world" (*Learning*, 34). This reflection on the learning of the Egyptians, passed to Moses, leads Bacon to King Solomon, whom Bacon believes "compile[d] a natural history of all verdure," citing the verse from 1 Kings about the cedar and the hyssop:

> *Salomon* became enabled, not only to write those excellent parables, or aphorisms concerning divine and moral philosophy; but also to compile a natural history of all verdure, from the cedar upon the mountain, to the moss upon the wall, (which is but a rudiment between putrefaction, and an herb) and also of all things, that breathe or move. Nay the same *Salomon* the King, although he excelled in the glory of treasure and magnificent buildings of shipping and navigation, of service and attendance, of fame and renown, and the like; yet he maketh no claim to any of those glories; but only to the glory of Inquisition of truth.
>
> (*Learning*, 35–6)[19]

Bacon gives the verse from 1 Kings expansive significance, not only transforming King Solomon into a natural historian but also implying a causal relationship between natural history, on the one hand, and advanced navigational technologies, wealth, and "renown" on the other. Bacon repeats this in Aphorism 129 of the 1620 *Novum Organum* (195) and again in his *New Atlantis*, posthumously published in 1627. Indeed, Bacon extends the myth of Solomonic natural history into the narrative arc of his fictional utopia, where we discover an ancient nation whose advanced navigational technology brings them wealth—both material and intellectual—and whose prosperity traces to Solomon. The governor of the House of Strangers claims to "have some parts of [King Solomon's] works which with you are lost; namely, that Natural History which he wrote, of all plants from the 'cedar of Libanus' to the 'moss that groweth out of the wall,' and of all 'things that have life and motion.'" Despite the difference in spelling, the narrator affirms that the House of Solomon was named after "the King of the Hebrews" in honor of his special association with natural history, its principal activity.[20]

The Solomon myth appears across the major texts of Bacon's *Great Instauration* and serves in each case as an apocryphal origin point for the study of nature, framed less as the discovery of new knowledge than as the recovery of ancient knowledge—more ancient than Plato and Aristotle. Mirroring the *Advancement of Learning*, where the reference to Solomon follows from a reflection on a lost body of Egyptian knowledge, the governor in the *New Atlantis* prefaces his account of the House of Solomon with a history of Atlantis that reaches back to Egyptian philosophy—passed on, in this case, not through Moses but Socrates. According to the governor of the House of Strangers, who narrates the history of Bensalem and the New World, the name "Atlantis" (which we learn refers to America) comes to Europe through an Egyptian, who shared his special knowledge of the lost civilization of Atlantis with Socrates. Socrates conveys the Egyptian's account of Atlantis in Plato's *Timaeus*—a "narration and description," according to the governor, "all poetical and fabulous: yet so much is true, that the said country of Atlantis, as well that of Peru [...] were mighty and proud kingdoms in arms, shipping, and riches" (*Atlantis*, 163). Bacon's chain of reasoning—from Moses to the Egyptians to Solomon and from Plato to the Egyptians to Solomon—would have been understood by contemporaries to refer to the *Hermetica* of Hermes Trismegistus and to the *prisca theologia*.

According to this theory, popularized by Marsilio Ficino but as ancient as Lactantius in the fourth century, Hermes Trismegistus, who was believed to be an ancient Egyptian priest, prophetically anticipated the revelations of Judeo-Christian scripture (*Hermetica*, xliii). Ficino credited Hermes as "the first author of theology"—"a single system of ancient theology, harmonious in every part"—who "foresaw the ruin of the old religion, the rise of the new faith, the coming of Christ" and more (*Hermetica*, xlviii). His complete translation of Plato's *Dialogues* into Latin appeared in 1484 under the patronage of Cosimo de' Medici's Florentine Platonic Academy, but his translation of Plato was interrupted at Cosimo's request to translate a newly recovered set of Hermetic texts. Ficino completed a Latin translation of the first 14 *logoi* of the philosophical *Hermetica* in 1463 (printed in 1471) under the title *Pimander*, the name of the god who reveals to Hermes the secrets of nature's creation in the first dialogue. Ficino's belief in a *prisca theologia* was informed by his work translating Plato; indeed, Hermes Trismegistus may be "the first author of theology," but Ficino believed theology "reached absolute perfection" in Plato (*Hermetica*, xlviii). Together, Ficino's *Pimander*, the Latin *Asclepius* (which was added to the Basel edition of Ficino's *Hermetica* in

1576), and *Platonic Theology* (printed in 1482) make up the most important sources of Hermeticism in early modern Europe.[21]

Renaissance authors erroneously believed that Hermes Trismegistus lived during the time of Moses, but they rightly identified the so-called philosophical *Hermetica* as soteriological and consistent, in many ways, with the teachings of the New Testament. Although André-Jean Festugière's position that the writings were gnostic in origin has been disproven, it remains true that they date to the time of Christ and that the texts available to Ficino and Bacon had been transformed by Gnosticism, which accounts for the similarity with Judeo-Christian scriptures that early moderns mistook for prophetic inspiration (*Hermetica*, xlvii).[22] The resemblances come through clearly in, for example, the first dialogue of Ficino's *Pimander*. Here, "Poimanders," whose name means "shepherd" (from ποιμήν, "shepherd," and ἀνήρ, "man"), inspires Hermes with a vision of the creation, which he then explains in terms of an origin myth: "Mind" or God the father "by speaking gave birth to a second mind, a craftsman," his son, the demiurge. The demiurge "crafted the seven governors" (seven planets) and separated Mind from Nature, i.e., the "weighty elements" that constitute material creation (*Hermetica*, 2). "Mind," after having spoken his son, the craftsman, into existence, "gave birth to a man like himself whom he loved as his own child":

> The man was most fair: he had the father's image; and god, who was really in love with his own form, bestowed on him all his craftworks. And after the man had observed what the craftsman had created with the father's help, he also wished to make some craftwork, and the father agreed to this. Entering the craftsman's sphere, where he was to have all authority, the man observed his brother's craftworks; the governors loved the man, and each gave a share of his own order. Learning well their essence and sharing in their nature, the man wished to break through the circumference of the circles.

And when he did so, Man saw his reflection in the waters of the earth. He loved the image of his own form and fell from heaven into Nature, who "embraced him, for they were lovers" (*Hermetica*, 3). Several elements of the creation of man in the *Pimander* inform sixteenth-century receptions of Hermeticism. As in the Christian narrative, man is created in the image of God, but the Hermetic Adam, brother to the craftsman or demiurge, is defined as a maker or creator rather than a beholder of creation: he has a knowledge of "all the craftworks," "wishes to make some craftwork," and operates in the same ways and in the same "sphere" as the demiurge. He has complete knowledge of the cosmos *qua* maker.[23] The Hermetic Adam, like the Mosaic Adam, also experiences a fall that results in death. But, unlike the Mosaic Adam, whose restoration depends on a savior, the Hermetic Adam's restoration depends on recovering his erstwhile powers of creative operation. Poimanders explains that man can resume his demiurgic status by relearning the secrets of the cosmos and, by re-ascending through the spheres of the seven guardians, "be made god" (*Hermetica*, 6).

These similarities between Christianity as Bacon understood it and elements of Ficino's *Hermetica* help to explain the plausibility and appeal of the *prisca theologia*—that is, of a single, unified theology that would bring later Christian teachings into alignment with other ancient sources. Bacon's *New Atlantis* registers this desire, formally, in the narrative prolepsis of the Bensalamites' conversion story. The governor of the House of Strangers recounts the discovery of an ark in the ocean—the Bensalemites, seeing a pillar of light rising out of the ocean, take to their ships in search of its source. They discover a cedar ark containing all the books of the Bible as it was known to Bacon (*Atlantis*, 159). As DeCook observes, the inhabitants of Bensalem have miraculously side-stepped the history of canonization and religious conflict suffered in Europe.[24] The narrative presupposes the existence of texts not yet written and debates not yet fought. In the same way that Bensalem miraculously has access to the lost natural history of King Solomon, so they have miraculous access to a post-Reformation edition of the Bible. The narrative prolepsis at the heart of Bacon's utopia gives formal resolution to a persistent historical tension between Judeo-Christian and Hermetic religious traditions. The effort to reconcile them ultimately failed. In 1614, Isaac Casaubon accurately dated Ficino's Hermetic texts to the era of Christ rather than Moses, challenging "the notion that pagan seers had predicted Christ's coming" (*Hermetica*, 1). Casaubon's humanist philology put an end to the *prisca theologia*, although, as Brian Copenhaver has argued, it took several decades to fade out completely. What remained in its wake was as important as any individual scientific discovery in the unfolding of modernity: historians of science generally recognize the salvific rhetoric of Bacon's *Great Instauration* as an important contribution to modern science.[25] The Hermetic promise of restoration through the mastery of nature and the role of Ficino's *Hermetica* in fomenting belief in a *prisca theologia* are integral to this legacy. But the way that Bacon reworks the Hermetic tradition is, like the story of the Bensalemites' conversion, proleptic in its reasoning. Bacon figures the Hermetic promise of restoration through operative power over nature metaphorically through the symbols of colonial expansion, especially in the *Novum Organum*. As in the *New Atlantis*, the discovery of the new world will make possible the recovery of ancient knowledge and thus the restoration of man to a prelapsarian state of pure power over nature.

The same formal prolepsis that turns the new world into a site of recovery rather than discovery in the *New Atlantis* informs Bacon's earlier works as well. And, as in the *New Atlantis*, Bacon enacts this transformation in and through the logic and metaphors of colonial expansion. Although scattered references to land surveys, navigation, and cartography in Bacon's earlier *Advancement of Learning* suggest that Bacon long conceived of knowledge as having a geography, the analogy grows in complexity in his later works, where the geography of learning culminates in a new world geography. The title *Novum Organum* is a strategic intertextual move on Bacon's part: Aristotle's logic was called, in Latin, the *Organum*; the *Novum Organum* is therefore a new logic, one meant to replace Aristotle's. But *organum* translates in English to "instrument," and in Bacon's colonial imaginary, the new "instrument" is figured as a mariner's compass:

> But as in earlier times when men only had the stars to sail by, they could indeed coast along the shores of the Old World or cross lesser and mediterranean seas; but before they could cross the oceans and discover the regions of the New World, the use of the mariner's compass, as a more trustworthy and certain guide, had first to be found out. By the same token, what has so far been discovered in the arts and sciences are of the sort that could have been discovered by use, meditation, observation and argument; for they lay pretty close to the senses and to common notions. But before we can get at the more distant and hidden aspects of nature, we are necessarily obliged to bring in means of bettering and perfecting the exercise and practice of the human mind and intellect.
> (*Organum*, 19–21)

Bacon's new logic will be to the discovery of new sciences what the mariner's compass was to the discovery of new worlds. The comparison between ancient and modern learning invokes a colonial geography to provide both a spatial and temporal framework for the classification of sciences: ancient learning is characterized by "the Old World," "shores," and "mediterranean seas," whereas the new science is figured in ocean crossings and "territories of the new world." The "Plan of the Work" also references this colonial foundation narrative as a structural metaphor for *The Great Instauration* as a whole. The survey of learning in Part I, for example, will "set a course along the coasts of the received sciences and arts" (*Organum*, 27), and, "after coasting past the old arts," Part II "will next make the human intellect ready to take to the high seas" (*Organum*, 29).

References to the mariner's compass and to transoceanic and intercontinental navigation constitute the metaphorical architecture of the new science in the 1620 *Great Instauration*. Another symbol of colonial expansion blazoned in the *Great Instauration*'s iconic frontispiece plays an equally important role in Bacon's geographic imaginary: the Pillars of Hercules. This image features a ship passing through the mythical pillars, which for the Greeks marked the limits of the known world. Just as the title of the *Novum Organum* gestures intertextually to Aristotle's *Organum*, so too do the Pillars of Hercules allude to Plato's *Timaeus*, where they mark the threshold of Atlantis. But the Pillars also had special resonance for the Spanish empire, since, as Delbourgo has argued, Bacon's frontispiece imitates Andrés García de Céspedes' earlier frontispiece to the *Regimiento de navegación* (1606)—which, like Bacon's work, features a ship passing through the Pillars of Hercules. Even before that, the Pillars and motto, "*plus ultra*," were adopted by Charles V of Spain in his personal emblem in 1516.[26] Bacon glosses their meaning in the *Advancement of Learning*, where the Pillars make up part of a more complex navigation metaphor: "For why should a few received authors stand up like Hercules Columns, beyond which, there should be no sailing or discovering, since we have so bright and benign a star as your Majesty to conduct and prosper us?" (*Learning*, 55). The preface to *The Great Instauration* similarly likens the "received arts" to "baleful pillars set up against the sciences, as men are not encouraged by the desire or hope of getting beyond them" (*Organum*, 11). The frontispiece, in other words, connects the foundational narratives of the new science formed in the *Advancement of Learning* and elaborated in the prefatory materials of *The Great Instauration* and *Novum Organum* to the *New Atlantis*, where they become explicitly colonial.

In both the *Novum Organum* and the *New Atlantis*, new world exploration enables the Hermetic fiction of the *prisca theologia*; the discovery of new worlds proleptically entails the recovery of ancient knowledge. The tools of colonial expansion that Bacon employs as symbols of the new science also came to symbolize the economic systems of extraction and exploitation that followed and maintained it. But it is worth emphasizing their original meaning within the symbolic economy of Bacon's *Great Instauration*: the mariner's compass, the Pillars of Hercules, the myth of Solomon, Daniel's prophecy—and, as I will argue in the next section, the ship that serves as the frontispiece's focal point—form a web of associations that connect Hermeticism and its emphasis on operation and the mechanical arts to the new world. Yates saw traces of Hermeticism in the utopias of Thomas More, Pico della Mirandola, and Bacon: "to me," she wrote, "it seems obvious that the New Atlantis has its roots in the Hermetic-Cabalist tradition of the Renaissance, though this is becoming rationalized in a seventeenth-century direction." Both Yates—for whom Bacon "draw[s] closer to some future Royal Society" than More or Mirandola—and Rossi believed that Bacon was a linchpin in the movement from Hermeticism to science, fomenting historical processes that would result in the formation of the Royal Society and the authorization of figures like Sloane.[27]

Hans Sloane and Baconian Natural History

The ship that features prominently on the frontispiece of Bacon's *Great Instauration* has received less attention than the Pillars of Hercules, but its design is significant. It has three masts, no forecastle, a high sterncastle, and a row of cannons positioned for use along its lower deck. Despite its occluded view, some of the ship's features are suggestive of a caravel—a small ship popular among early explorers for its speed and maneuverability.[28] The ship on Bacon's frontispiece has the lighter build of a caravel but also, because of its menacing row of canons, the warfighting capacity of a galleon. The galleon was in popular use by the late sixteenth century as both a merchant vessel and warship because of its larger size, stability, and load capacity, which allowed it to carry both goods and munitions. Bacon's ship, with its combined elements of the caravel and galleon, can be read emblematically as a symbol of the combined forces of exploration and exploitation that fueled early modern Spain's colonial economy and would eventually do the same for England.[29] In this section, I argue that Sloane's specimen plate displaying a Man-of-War jellyfish, Spanish coins, and part of a sunken Spanish galleon, together form a visual intertext with the ship on Bacon's frontispiece. Sloane's specimen plate gestures both toward Spain's colonial past and toward the Baconian past; the early modern origins of Sloane's colonial project trace not just to the mercantilist hegemony of Spain, whose fall from power is blazoned forth in the coral-encrusted remains of a wrecked Spanish galleon, but also to the Hermetic soteriology of Bacon's *Great Instauration*. If Bacon believed that the recovery of Solomonic natural history and the restoration of the fallen, Hermetic Adam could be proleptically fulfilled by colonial expansion into the New World, Sloane's engraving positions the natural-historical specimen as the heir and agent of Bacon's vision.

Sloane's *Voyage* includes several texts in two volumes. The first volume, published in 1707, contains a preface and lengthy introduction that provides a chorography of Jamaica, including its history, geography, weather, inhabitants, and government. The introduction is followed by the *Voyage to Jamaica*, a travel narrative, and finally the *Natural History of Jamaica*, a collection of specimen descriptions and series of numbered specimen plates. The second volume, published almost twenty years later in 1725, includes another introduction, the *Voyage from Jamaica*, another *Natural History of Jamaica*, errata, index, and more numbered specimen plates. In total, the *Voyage* contains 274 specimen plates, but the specimen plate featuring the jellyfish, coins, and timber belong to a separate group of eleven plates that divide the introduction from the travel narrative in volume one. Kriz notes that while "the placement of these eleven plates—four in volume one, seven in volume two—varies from copy to copy," nevertheless, "in all cases that I know of, they precede the 274 sequentially numbered plates."[30]

Initially, viewers might not recognize the coins and pieces of shipwreck, which are so encrusted with coral that they look like natural specimens rather than artifacts as such. In this respect, they complement Sloane's selection of the Man-of-War jellyfish, so named because of its resemblance to a caravel. In fact, Sloane emphasizes the resemblance by labeling it "A Carvell." Elsewhere in volume one, Sloane describes his first encounter with one at sea: it is "what the seamen call a *Caravel* or *Portuguese* Man of War," whose "stinging" is "much worse than Nettles, whence it is by some reckoned Poisonous." Sloane's repeated emphasis on its "poisonous" sting accounts for its association not just with a ship but a warship. Like a Portuguese Man-of-War, the sting of Sloane's "Carvell" is uniquely "violent": it "burn[s] more violently than those [jellyfish] of the North-sea," and they "suck themselves so close to the skin that they [do] raise blisters, and cause sometimes St. *Antony's* Fire" (1:7-8). But also like a caravel, they are capable of skilled maneuvering in the water: "Seamen do affirm that they have great skill in sailing, managing their bladder or sail with judgement as may be most for their purpose, according to their different winds and courses." Although Sloane denies the sentience implied by this belief—the seamen "[allow] them more reason, than I, at present, am willing to do of life"—the anecdote reproduces, in the figure of the jellyfish, the logic of Bacon's caravel-galleon (1:8).

Formally, the selection of objects—a natural specimen that looks like a ship and artifacts that look like natural specimens—and their carefully balanced arrangement on the page create a chiastic tension between nature and artifice that lends aesthetic unity to the composition. But these Spanish artifacts also make a historical argument about the rise and fall of empires. As Kriz argues, this specimen plate represents "the Spanish presence on the island and dominance over the entire region in the sixteenth and seventeenth centuries"—it summons the "Spanish ghosts" of Jamaica's past.[31] The coins and timber were recovered by Sloane's patron, the Duke of Albemarle, from a Spanish galleon that sank in 1659. Sloane recounts the venture in "The Introduction" to volume one:

> This wreck had been a Spanish galleon lost on these shoals, near the *Abreojos* or *Handkercher* shoals, to the north-east of *Hispaniola*, about the year 1659 bound for Spain. [...] in most parts there is such a vegetation of coralline matter out of the sea-water, as that the bottom of the sea is incrustated with it, and the wrecks hid by them. The Pieces of Eight in the silver-wreck above mentioned, that was lost in 1659 were covered with this matter about a quarter of an inch thick, and I have a piece of the timber of the ship, with an iron bolt in it, grown over with the *Corallium asperum candicans adulterinum* J.B. and some of the Pieces of Eight incrustated, others almost covered with *Astroites*. Those underneath were corroded with the sea water, so that many of them stuck together.
>
> (1:lxxx)

Although separated by over fifty pages, both the Latin inscriptions and this passage from the introduction identify the plate's "incrustated" artifacts as specifically Spanish. The introduction to volume one opens by narrating the explorations of Christopher Columbus, the settling of Jamaica by the Spanish, their abandonment of it, and the English possession of it in 1655. Indeed, the Spanish feature prominently throughout the introduction as the island's first colonizers and as violent aggressors in the region generally. But the plate's specimens also associate the Spanish with ruin and decay. The plate gestures toward the decline of Spanish colonial power—visible everywhere in the ruins of abandoned Spanish plantations observed across the island and recorded by Sloane—and the hoped-for rise of the English. It is worth emphasizing the importance of Sloane's shift into English when labeling the jellyfish. Although calling it elsewhere a Portuguese Man-of-War and placing it next to the remains of a sunken Spanish galleon invokes the once-great power of the Spanish and Portuguese during the earliest period of exploration, the shift from Latin ("*Urtica marina soluta, purpurea, oblonga, cirrhis longissimis*," or "marine nettle, loose, purple, oblong, with very long tendrils") into English ("A Carvell") suggests that it is an English warship that Sloane positions against the ruins of Spanish colonial empire.

As this reading indicates, the cross-referencing required to interpret Sloane's engravings transforms each specimen into an emblem. Iannini, expanding on Kriz's emblematic readings of Sloane's specimens, draws a connection between early modern emblems and the religious providentialism guiding early natural history specimens. Richard Ligon's *True and Exact History of the Island of Barbados* (1657) takes, for example, the cross visible in a bisected banana as a providential reproach to European colonists for failing to evangelize to enslaved and indigenous communities.[32] Although Iannini reads Sloane's later depiction of the jellyfish and wreckage as having transformed Ligon's religiously-inflected specimen-emblems into instruments of financial speculation, the composition's implied trajectory from Spanish to English hegemony does appear to "retain the basic function of the emblem as an orienting device for discerning providential patterns."[33] Indeed, in this case, the providential arc of English imperial hegemony depends upon the commodification of Jamaican natural specimens.

What makes the providential reading of this particular specimen plate possible is its chiastic juxtaposition of natural and artificial specimens. The commodification of Caribbean natural specimens—from sugarcane to rum to cotton to cacao—depended on the imbrication of the mechanical arts and natural history. In other words, the imbrication of nature and artifice was mutually constitutive of the new science.[34] Although Sloane was not alone in including descriptions of machinery in texts otherwise dedicated to travel narrative and natural history—Ligon's *History* includes a detailed description of a sugar mill—it is worth noting the Baconianism of the convention. In Aphorism 98 of the *Novum Organum*, Bacon critiques the natural histories of his day for including only "natural species."[35] To produce new technologies, not just knowledge for its own sake, a natural history must include with its "variety of natural species" also the "experiments of the mechanical arts":

> For the *raison d'être* of a natural history drawn up for its own sake is one thing, but one compiled systematically to inform the intellect for the building up of philosophy is quite another. And these two histories differ in many respects but above all in this: that the first takes in the variety of natural species but not the experiments of the mechanical arts.
>
> (*Organum*, 157)

Sloane takes up Bacon's call by, for example, including a detailed engraving of a cotton gin among the 274 specimen plates in the *Natural History of Jamaica*. He also follows Bacon by dedicating the plate depicting the jellyfish, coins, and timber to the mechanical art that served as a governing conceit of the *Great Instauration*: navigation. Bacon's licensing of the mechanical arts as both useful and necessary in the pursuit of natural history—a view shaped by the Hermetic emphasis on operation—finds an important afterlife in Sloane.

I have suggested the ways in which Sloane's jellyfish reproduces the emblematic argument of Bacon's ship: combining elements of both caravel and galleon, Bacon's ship and Sloane's "Carvell" blazon forth the twin colonial principles of exploration and exploitation. But Sloane ominously adds a final layer of meaning to the "Carvell" when, in his account of the caravel in volume one of the *Natural History of Jamaica*, he cites Richard Hakluyt's name for it: "a ship of *Guinea*" (1:7). As Iannini reminds us, a "ship of *Guinea*" was a slave ship.[36] Embedded in the specimen's many names is a declaration of England's nascent role in slave trafficking as an engine of imperial power. In this respect, Sloane clarifies a latent tension that goes unresolved in Bacon's new-world vision of the new science. In his essay "On Plantations," Bacon opens with the assertion that "I like a plantation in a pure soil; that is, where people are not *displanted*, to the end, to *plant* in others. For else, it is rather an extirpation, then a plantation."[37] He closes the essay, however, with a warning: "If you plant, where savages are, do not only entertain them with trifles, and jingles; but use them justly, and graciously."[38] In other words, the opening of the essay imagines a new world without people, but ends by giving advice for how to deal with the people already living there. Bacon acknowledges the contradiction in his use of conditionals: he "likes" an unpeopled land but "if" you find people there, deal fairly with them.

By the publication of Sloane's *Voyage*, the "extirpation" that Bacon warns of was in full force—Jamaica was, as Vincent Brown has argued, "a death trap."[39] Charles II renewed the Royal Society's charter in 1663—the same year he granted a charter to the Company of Royal Adventurers Trading into Africa, one of the earliest English companies to engage in the slave trade. Mark Govier has traced the entanglement of science and slavery through the history of the Royal Society and its investments in the slave trade. Not only were fellows of the Royal Society, including its leadership, personally involved in the slave trade as owners, investors, and enslavers, the Royal Society also invested its own funds in such ventures as the East India Company and the Royal African Company. The coincident rechartering of both the Royal Society and the Royal Adventurers, the shared membership across colonial government and the Royal Society, and the direct financial investment in the slave trade from which the Royal Society benefited implicate the origins of modern science in the origins of colonial slavery.[40] Sloane's *Voyage* exemplifies the incremental transformation of the Hermetic myth of man's power over nature into the extractive sciences of early capitalism. In this transformation, the relationship between the mechanical arts and the natural world becomes codified in specimens-turned-commodities. In the course of this process, we witness also the foreclosure of humane forms of encounter: whereas Bacon, writing at the turn of the seventeenth century, could still acknowledge the greatness of new world civilizations and imagine the reuniting of old and new worlds as part of a larger redemptive narrative of human flourishing, by the time of Sloane the internal contradictions in Bacon—who cannot reconcile the presence of other "people" without transforming them ultimately into "savages"—have been resolved in the direction of colonialist catastrophe, with dire human and environmental consequences. If Bacon's Solomon "compile[d] a natural History of all verdure," "excelled in the glory of treasure and magnificent buildings of shipping and Navigation," and "glory[d]" in the "Inquisition of truth," it is the specimen-emblem of Sloane's *Voyage*—the Man-of-War jellyfish that is also "a Carvell," and ultimately a "Ship of *Guinea*"—that declares itself the true *novum organum* or "new instrument" of Bacon's *Great Instauration* and Sloane its Solomon.

In past generations, modern preconceptions about what counted as science led to the exclusion of other important intellectual traditions—from alchemy to astrology to cabala to Hermeticism—from the history of science. Mid-century scholars like Walker, Yates, and Rossi succeeded in opening the history of science up to intellectual traditions not obviously scientific in nature but formative to its development. Our understanding of the history of religion would benefit from a similar openness to what I have called here "minor religious traditions." The influence of Hermeticism on the interpretation of scripture—in this case, the interpretation of 1 Kings 4:33—is as important to the history of religion as to the history of science.

Notes

1 Sloane, *Voyage to Jamaica*, 1:65. Hereafter cited parenthetically in the text by volume and page number.

2 Graham Rees, "Introduction," in Francis Bacon, *Instauratio Magna Part II: Novum Organum*, ed. Rees and Wakely, xlvii. Hereafter cited parenthetically in the text as *Organum* followed by page numbers.

3 Susan Bruce, "Introduction" to Francis Bacon in *Three Early Modern Utopias,* ed. Bruce, xxxi. Anthony Grafton has argued that although "Salomon's House hardly resembles the schools, learned academies, and universities of the sixteenth century," Bacon may have taken inspiration from Matthias Flacius Illyricus and his Magdeburg Institute (*Worlds Made by Words*, 100). For Bacon's relationship to the Magdeburg Institute and the influence of the *Magdeburg Centuries* on Bacon's *New Atlantis*, see Grafton, "Where Was Salomon's House?" (in *Worlds Made By Words*, 98–113).

4 Michael Kiernan, "Commentary," in Bacon, *Advancement of Learning*, ed. Kiernan, 239. All passages from this book by Bacon are hereafter cited parenthetically in the text as *Learning* followed by page number.

5 For an account of Bacon's reception see Rees, "Introduction" in Bacon, *Organum* (xxii–xviii); see also Thomas Kuhn, "Mathematical vs. Experimental Traditions," 2.

6 Daniel 12:4 (King James Version).

7 Delbourgo, *Collecting the World*, 28. As Rees explains, Bacon understood Daniel's prophecy to mean "that Providence has ordained that the voyages of exploration and scientific advancement are destined to be coeval" (*Organum*, lxii).

8 For Sloane's Baconianism, see Delbourgo, *Collecting the World*, 25–9. For the separation of Sloane's herbarium from the rest of his collections as part of his scientific legacy, see Delbourgo, "Sir Hans Sloane's Milk Chocolate," 72–3 and 92–3; for Sloane and the history of the museum, see also MacGregor (ed.) *Sir Hans Sloane*.

9 Among the earliest articulations of Bacon's supposed secularism was Richard F. Jones. See *Ancients and Moderns*, 63. More recent echoes of Jones can be found in Brad Gregory, *The Unintended Reformation*, 57; and Markku Peltonen, "Introduction," *Cambridge Companion to Bacon*, ed. Peltonen, 19. See Cooper, "Francis Bacon's Idols," 332.

10 For the term "Hermetic Adam," in contrast to the "Mosaic Adam," see Yates, "The Hermetic Tradition," 257.

11 I borrow the term "specimen-emblem" from Iannini, who coins the term "specimen-as-emblem." See *Fatal Revolutions*, 26.

12 Poole, "God's Game of Hide and Seek," 118.

13 DeCook, "The Ark and Immediate Revelation," 105–6.

14 See, for example, D. P. Walker, *The Ancient Theology*; Rossi, *Francis Bacon*; and Yates, *Giordano Bruno*.

15 Yates, "The Hermetic Tradition," 255.

16 Whereas Yates emphasizes continuity with the past, Rossi emphasizes Bacon's modernity. Nevertheless, Rossi always affirmed "Bacon's profound indebtedness to the magico-alchemical tradition and to the Renaissance concept of magic" ("Hermeticism and Rationality," 256).

17 Walker, "The *Prisca Theologia* in France."

18 See *Hermetica*, ed. Copenhaver, l. Hereafter cited parenthetically in the text as Hermetica, followed by page number.

19 See also Poole, "God's Game of Hide-and-Seek," 124–5.

20 Bacon, *The New Atlantis*, in *Three Early Modern Utopias,* ed. Bruce, 167. Hereafter cited parenthetically in the text by page number as *Atlantis*.

21 For the dates of translations and publications, see Scott, *Hermetica*, 31; Yates, *Giordano Bruno*, 12–13; and Copenhaver *Hermetica*, xli.

22 See also Yates, *Giordano Bruno*, 12–18.

23 On the distinction between beholding and making, see Antonio Pérez-Ramos, *Maker's Knowledge*, 48–62.

24 DeCook, "The Ark and Immediate Revelation," 114.

25 See, for example, Debapriya Sarkar, *Possible Knowledge*.

26 See Delbourgo, *Collecting the World*, 23; Margery Corbett and Ronald Lightbown, *Comely Frontispiece*, 186; and Earl Rosenthal, "Columnar Device," 204.

27 Yates, "The Hermetic Tradition," 265.

28 Guilmartin, *Galleons and Galleys*, 86–97. In the fifteenth century, the Portuguese used the caravel as a warship, but cannons were placed on the main deck and often they could only carry a single canon, thus "Caravels retained their utility throughout the sixteenth century, but were never first-line warships in European waters" (89). Similarly, Carracks were used as warships, but were bigger than the ship on Bacon's frontispiece, and were not used for transoceanic navigation because their bulk made them harder to defend.

29 Ibid., 158–64. The main difference between a caravel and a galleon—other than one interesting fact, that "the English applied the word [galleon] more often to foreign vessels than their own" (158)—is their high fore- and stern-castles, "giving the galleon its characteristic crescent shape when viewed from the side" (158). Neither of the two ships represented in Bacon's frontispiece has the crescent shape of a galleon because they appear to lack a forecastle, but the heavy ordinance on the lower deck mark the ship as a galleon: "galleons were the first sailing vessels routinely capable of transoceanic navigation that could effectively bring heavy guns to bear offensively" (159). Guilmartin also describes the "race-built galleon" popular under Elizabeth I, but notes that "race-built galleons were not true transoceanic warships" and were used, instead, for domestic defense (160–1).

30 Kriz, *Slavery, Sugar, and the Culture of Refinement*, 15.

31 Ibid., 19–20.

32 Iannini, *Fatal Revolutions*, 26–8.

33 Ibid., 25. For Iannini's reading of Sloane's jellyfish, see 69–74.

34 Although as Kriz notes, "the next major American natural history to appear after Sloane's, Mark Catesby's two-volume *The Natural History of Carolina, Florida, and the Bahama Islands* (1731–43), eschewed illustrations of artifacts as a means of engaging the curiosity and interest of readers; West Indian natural histories produced later in the century followed Catesby's example" (*Slavery, Sugar, and the Culture of Refinement*, 35).

35 This critique may refer to Pliny and Aristotle, whose natural histories Bacon names elsewhere. But it may be directed at Conrad Gesner, whose species-centric *Historia animalium* (1551–8) William Ashworth describes as "the most widely read of all Renaissance natural histories" ("Emblematic natural history," 17).

36 Iannini, *Fatal Revolutions*, 71.

37 Bacon, *Essays*, ed. Kiernan, 106.

38 Ibid., 108.

39 Brown, *The Reaper's Garden*, 13.

40 Govier, "The Royal Society," 204–9.

6

Counting Time: Then and Now

Jonathan Sawday

Who saw
When this creation was? Remember'st thou
Thy making, while the maker gave thee being?
We know no time when we were not as now;
Know none before us, self-begot, self-raised
By our own quickening power.
—John Milton, *Paradise Lost* (1667) V. 856–61

Thou hast ordered all things in measure and number and weight.
—Wisdom 11:20 (King James Version)

Introduction: The Creation Museum

Situated in northern Kentucky and owned and operated by the Christian fundamentalist organization "Answers in Genesis" (AiG), the Creation Museum promotes the pseudoscience of "Young Earth Creationism." It has been argued that the Creation Museum is undeserving of the title "museum" in the first place, since it is "based on a conscious conflation of science and faith, two realms that can certainly co-exist and complement one another but that are not by their very nature able to provide verification for the other."[1] Nevertheless, following its opening in May 2007, the Creation Museum proved to be popular with a large swathe of the American public, welcoming over 400,000 visitors in its first year of operation.[2]

The Creation Museum is dedicated to the proposition that the opening chapter of Genesis, recounting how, in the space of six terrestrial days, "God created the heaven and the Earth" together with night and day, the sun and the moon, the oceans and the dry land, plants and herbs, "every living creature that moveth," and (finally) "man in our image, after our likeness" (Gen. 1:1-20, King James Version), is an historically and scientifically accurate record of the world's beginning. The Creation Museum is also committed to the view that the divine *fiat* of creation was pronounced in the relatively recent past, around 6,000 years ago. Flying in the face of modern geology, cosmology, paleontology, biology, zoology, and a battery of radiometric dating methods, which, cumulatively, provide evidence for an ancient Earth formed *circa* 4.54 billion years ago, the Creation Museum also sets out to refute what it terms "the naturalistic evolutionary worldview."[3] Relying on text-heavy displays and diagrams, animatronic figures, and extravagant dioramas, the museum shows dinosaurs and modern humans happily coexisting with one another in a recreation of the Garden of Eden.[4] With its strapline, "Prepare to Believe," the Creation Museum offers "an impressive and sophisticated visual argument on behalf of young Earth creationism and a highly politicized fundamentalism."[5]

Archbishop Ussher and Biblical Chronology

Although AiG is a modern foundation, its intellectual roots may be traced back to the mid-seventeenth century. In 1650 the archbishop of Armagh, James Ussher, published his *Annales Veteris Testamenti* (Annals of the Old Testament) in which he famously (or notoriously) announced that the world began at some point shortly before Sunday, October 23, 4004 BCE. As reported in 2016, copies of Sebastian Adams's *Synchronological Chart or Map of History*, first published in 1871 and based in large part on Ussher's dating system, were on sale in the Creation Museum's bookshop, where they were described as "perfect for homeschool settings" and "Sunday school walls."[6] Today, both the Adams chart and an edited (and modernized) version of Ussher's *Annals of the Old Testament* may be purchased via the AiG website. The website describes how Ussher's seventeenth-century *Annals* were "mocked when evolutionary theories became vogue in the 19th century," continuing:

> Archbishop James Ussher was a well-respected and important scholar of ancient history during the early 17th century. A meticulous researcher and compiler, Ussher correlated the events in Genesis with secular human history [...]. *The Annals of the World* was so highly regarded for its preciseness that its timeline was included in the margins of many King

> James Bibles throughout the 18th, 19th and 20th centuries, reminding readers that the Earth is only around 6,000 years old.

But today, as the website also explains, "few Bible publishers include Ussher's dates because they conflict with the popular belief in 'millions of years,' which is based on evolutionary assumptions." Thus, Ussher's "sound scholarship" was banished from Bibles, reflecting "the church's compromise with Darwinism."[7]

This account, although accurate in its representation of Ussher's seventeenth-century reputation as a scholar of ancient history, is misinformed when it comes to explaining his fall from scientific grace. The nineteenth-century demolition of Ussher's time scheme was initiated not by skeptical followers of Charles Darwin but by an American evangelical scholar, William Henry Green, who had originally set out to *defend* the Genesis account of creation. In 1863, Green published *The Pentateuch Vindicated from the Aspersions of Bishop Colenso*, in which he sought to refute the views of John William Colenso, the Anglican Bishop of Natal. In his earlier work on *The Pentateuch and the Book of Joshua Critically Examined* (1862), Colenso had drawn on Charles Lyell's *Elements of Geology* (1838) to argue "that a *Universal* Deluge, such as the Bible manifestly speaks of, could not possibly have taken place in the way described in the Book of Genesis."[8] Keen to dismiss Colenso's doubts, Green, an Old Testament scholar at Princeton Theological Seminary, nevertheless had to concede that there was room for uncertainty. His hesitation was precipitated not by the vastly expanded notion of geological time heralded by the publication of James Hutton's *Theory of the Earth* (1788) and extended in Lyell's *Principles of Geology* (1830–3), but by the Bible's own method of counting time, derived from genealogical records:

> But it must not be forgotten that there is an element of uncertainty in a computation of time which rests upon genealogies, as the sacred chronology so largely does. Who is to certify us that the ante-diluvian and ante-Abrahamic genealogies have not been condensed in the same manner as the post-Abrahamic? If Matthew omitted names from the ancestry of our Lord … may not Moses have done the same in order to bring out seven generations from Adam to Enoch, and ten from Adam to Noah?

The Bible, just like any text, probably contained gaps or compressions, so that calculating a chronology based on the generations of the descendants of Eve and Adam could quite possibly yield an erroneous result; as Green wrote, "a select and partial register of ante-Abrahamic names has been mistaken for a complete one."[9] This dispute was not between godless Darwinians and believers in biblical inerrancy but between two wings of late nineteenth-century Anglo-American Protestantism, in which one side, driven by a commitment to the truth of scripture, had begun to sense, nevertheless, a flaw not in the method of calculation but, rather, in the data upon which the calculations were based.

Ussher's precise dating of the moment of the world's creation seems to have prompted very little comment in England until the later seventeenth century. John Quarles's fulsome funeral elegy on Ussher, composed shortly after the archbishop's death in March 1655, made no mention of his chronological work, concentrating, instead, on his fierce anti-Catholicism, as did another (anonymous) funeral elegy published in 1656.[10] The sermon preached at Ussher's funeral by his former chaplain, Nicholas Bernard, and revised and published as a short biography in 1656, barely glanced at the archbishop's chronology, although it was full of praise for his extraordinary learning and his defense of the reformed church in Ireland.[11] Not until 1686, when Richard Parr's *The Life of the Most Reverend Father in God, James Ussher* appeared, did his chronology warrant any sustained attention. But the dating of creation was entirely ignored by Parr, who, instead, praised Ussher for the precision of his calendrical "Rules whereby to find out the *Cycles* of the *Sun*, and Moon, and *Easter* for ever" and his understanding of "Celestial Motions, according to the Ancient *Greek* Astronomers" together with his "*Astronomical*, as well as Chronological Learning."[12]

Ussher's fame as the calculator of the moment of the world's commencement did not begin until well after his death. In 1701, William Lloyd, the Anglican Bishop of Worcester, supervised the publication of a folio edition of the King James Bible, in which, for the first time, the archbishop's chronology was printed in the margins of the text.[13] Thereafter (as the AiG website correctly reports), many imprints of the King James Bible offered Ussher's chronology, such as the tables and explication entitled "Bases of Scripture Chronology," compiled by the British Baptist minister, educator, and biblical scholar Samuel Gosnell Green.[14] Green's chronology drew on two main sources: Ussher's mid-seventeenth-century chronological calculations and William Hales's influential revision of Ussher's chronology, which was published in three volumes as *A New Analysis of Chronology* (1809–12)—a work that had taken him twenty years of labor.[15] Despite his efforts, Hales could only push back the date of creation by just over 1,400 years to offer a revised date of 5411 BCE as opposed to Ussher's 4004 BCE.

In his "Bases of Scripture Chronology," Green maintained that establishing a chronology from the "Creation of Man to the Deluge" was an "easy calculation" for Ussher to perform, using Genesis 5—"the book of the generations of Adam"—as the starting point.[16] In fact (as we shall see), it was by no means easy. Moreover, although today Ussher's reputation stands (or falls) on his notorious dating of the first day of creation, establishing such a date was not the primary concern of his *Annales*. Rather, Ussher was attempting to present a "comprehensive chronological synthesis of all known historical knowledge, biblical and classical, down to just after the fall of Jerusalem in AD 70."[17] To this end, Greek and Roman history was just as important to Ussher as biblical and Jewish history.

Today, however, Ussher's broader chronological work is largely forgotten. Rather, he is almost universally derided for his dating of the world's beginning. As the evolutionary biologist Stephen Jay Gould has remarked (in a quixotic defense not of Ussher's conclusions, but of his methods): "to this day, one can scarcely find a textbook in introductory geology that does not take a swipe at Ussher's date as the opening comment in an obligatory page or two on older concepts of the Earth's age." And Gould goes on to heap scorn not on Ussher, whose work he defends as "honorable" in terms of seventeenth-century systems of thought, but on the archbishop's latter-day detractors who have invested the Ussherian scheme with additional layers of simple-minded literalism. Quoting one (unnamed) modern geology textbook, which claims that Ussher reckoned the world to have been created "at 9:00 A.M., October 26, 4004 B.C. (presumably Greenwich Mean time!)," Gould writes that this observation, "with its feeble joke about Greenwich time," is itself incorrect, since what Ussher "actually said" was "October 23 at noon."[18]

In fact, Gould, too, was incorrect. Ussher "actually said" no such thing. In his *Annales veteris testamenti*, Ussher began his account of biblical chronology as follows: "*In Principio creavit DEUS Coelum & Terram* [Genes 1.1.] *quod temporis principium (juxta nostra Chronologiam) incidit in noctis illium initium, quae XXIII. diem Octobris praecessit, in anno Periodi Julianae 710.*"[19] When the work was published in an English version in 1658, this opening sentence was rendered as: "In the beginning God created Heaven and Earth, *Gen. 1. v. 1.* Which beginning of time, according to our Chronology, fell upon the entrance of the night preceding the twenty third day of *Octob*[*er*] in the year of the Julian Calendar, 710."[20] At no point does Ussher venture a precise time of day for creation. But from the phrase "upon the entrance of the night preceding the twenty third day of Octob[er]," we can infer that, in Ussher's reckoning, not just the world but time itself began around sunset (as we would measure it) on October 22, 4004 BCE, except, of course, that the sun had yet to be created. The famous date was to be found not in the narrative of biblical chronology, but in the chronological columns that bracketed the text block in the margins of each page in either the Latin or English versions of the

Annales. In those columns, we find three distinct dates given for each datable event recorded in the Old Testament: "*Anno Mundi*" (Year of the World), "*Anno Periodi Julianae*" (Year of the Julian Period), and "*Anno ante aeram Christianam*" (Year before the Christian era). Thus, according to this scheme, time (and the world) began at the start of Year 1 of *Anno Mundi*, which was also the beginning of year 710 of the Julian Period, and 4004 years before the birth of Christ. But if time did not yet exist, then how do we account for the 709 years of the "Julian Period" which, according to Ussher's calculations, had already elapsed before the creation?

Julian Time

Ussher's "Julian Period" was the invention of the Franco-Italian Calvinist scholar Joseph Justus Scaliger, who, in his 1583 *De emendatione temporum* (On the Emendation of Time), had proposed a method of calculating elapsed time, where the customary Western BC/AD (now BCE/CE) date was either unknown, the product of a different calendrical system, or recorded incorrectly. The Julian system was devised to liberate the measurement of time from any one religious point of view. As Anthony Grafton writes, Scaliger's method

> aims not to find a moral order in the past, but simply to reconstruct that past; it employs not merely the one divinely-inspired source, but all sources. Most important, it is based on a philological method which applies equally to Hebrews and to Greeks, to the Bible and to the ancient historians.[21]

By offering "a kind of reference timeline independent of any real history," the Julian scheme allowed for the computation of time relatively unburdened by the cultural assumptions inherent within competing Old World calendrical systems, whether ancient or modern: Christian, Roman, Hebrew, Egyptian, Zoroastrian, Babylonian, Islamic, Hellenic, or Hindu.[22] It also allowed time to be expressed as a single flow of units in one direction, unlike the clumsy BC/AD system, which requires counting both backwards and forwards from the assumed date of the birth of Christ.

Far from being imbued with biblical certainty, for many of the contemporaries of Scaliger and (later) Ussher, the nativity of the world was a matter fraught with mystery. In the two influential hexaëmeral poems of

Guillaume de Salluste, Sieur du Bartas, published as *La Semaine ou Création du mond* (1578) and *La Seconde Semaine* (1584), and translated into English as *The Divine Weeks* (1604), the beginning of time and the world was beyond human reason. So, the instant when "Time first became" was also a "Time confused" since any means of marking its passage—"years … months … weeks … days"—before the commencement of "the ordered Dance unto the Stars assigned" was lacking, while creation itself was the work of a deity who was "Immutable, immortal, infinite / Incomprehensible."[23] Similarly, glossing the phrase "*In principio*" ("In the beginning") with which Genesis opens, Sir Walter Raleigh's unfinished *History of the World*, first published in 1614, asserts that any attempt to impose a human timescale on the moment when "things began to be" is doomed to failure. For, Raleigh wrote, "the word *beginning* (in which the Hebrews seek some hidden mystery, and which in the *Jews Targum* is converted by the word *Sapientia*) cannot be referred to succession of time nor to order."[24] For the polymathic Robert Burton, in his syncretization of classical and modern knowledge which was *The Anatomy of Melancholy* (1621), the question of beginnings posed, in the end, an insoluble series of riddles:

> Some again, curious fantastics, will know more than this, and inquire with Epicurus, what God did before the world was made? Was he idle? Where did he bide? What did he make the world of? Why did he then make it, and not before? If he made it new, or to have an end, how is he unchangeable, infinite, &c?

The pursuit of the prehistoric past was a fruitless undertaking. So, Burton observed of the "history and records of former times" that "God in his providence, to check our presumptuous inquisition, wraps up all things in uncertainty, bars us from long antiquity, and bounds our search within the compass of some few ages."[25]

If only the immediate past (which, for Burton, included pagan antiquity) could be knowable, then all that was left was conjecture. In Book VI of his *Pseudodoxia Epidemica* or "Enquiries into very many received Tenets and Commonly Presumed Truths," first issued in 1646, Sir Thomas Browne also explored the question of the age of the Earth. His essay began by plainly setting forth his conclusions:

> Concerning the World and its temporal circumscriptions, who ever shall strictly examine both extremes, will easily perceive there is not only

> obscurity in its end but in its beginning; that as its period [i.e., ending] is inscrutable, so is its nativity indeterminable.[26]

Adopting his customary method of undertaking a survey of authorities stretching back to classical antiquity, glancing at Jewish and Arabic sources, as well as the "Chronology of the moderns," including Scaliger, comparing different translations of the sacred texts, and sifting and sorting the reliable from the unreliable, Browne found only confusion and contradiction. So, he recorded a plethora of different dates for the beginning of the world: 3947 BCE, 5664 BCE, 5515 BCE, 5509 BCE, and so on. All these various dates are, however, within the "range" (as we might now say) proposed by Ussher, underlining the sense that Browne, too, was thinking of time in a similar dimension. However, this "difference and wide dissent of Men's opinions" led Browne to the gloomy conclusion that the calculation was "beyond the Arithmetic of any but God," so that "He that undertakes the knowledge thereof, forgets his own beginning."[27]

That was not, however, how Scaliger had approached the problem. In Scaliger's system, time is computed by combining three distinct calendrical cycles, two of which are based on astronomical data—the 19-year lunar cycle and the 28-year solar cycle—and one on an ancient time division: the 15-year Roman Indiction Cycle, which measured the time between taxation assessments throughout the Roman empire and which continued to be used as a dating method in late antiquity and in medieval Europe.[28] The product of these three cycles (19 × 28 × 15) produces a Julian Period of 7,980 years, which was held to commence when all three cycles began in the same year. Thus, within the current Julian Period, the year 2025 CE can be recorded as Julian Year 6738, in a cycle of 7,980 years which commenced (using the customary Western dating method) on January 1, 4713 BCE and which will end on December 31, 3267 CE.

Using this system, it is also possible to record a Julian Day number not just for every day, but even fractions of a day, starting at Day 1 which began at midnight on January 1, 4713 BCE. For example, the sentence you are now reading was written on December 7, 2024 CE, at 2.59 p.m. (US Central Time) or 20:59 (UTC). This moment can be expressed in Julian Days as a single numeric: 2,460,652.37 (i.e., just over a third of a day into the 2,460,653rd day after midnight, January 1, 4713 BCE).[29] The advantage of the Julian system is that it allows for the calculation of elapsed time between events in the distant past as well as the distant future, while also recording time as a single numeric, rather than as a cumbersome combination of words and numerals, or the confusingly ambiguous (modern) conventions of expressing the date numerically in different cultures: DD/MM/YYYY as opposed to

MM/DD/YYYY, or YYYY/MM/DD, the latter being (today) the scientifically accepted International Standard Date Notation.[30] It was the apparent precision of the Julian system which helped Ussher to assign not merely years and months to events recorded in the Bible, but days and even the approximate time of day on which events might have occurred, and which, in turn, has occasioned the modern derision for the apparent literal-mindedness of the seventeenth-century chronologer.

Despite the pessimism of Burton and Browne, in England the Julian system had been used some years prior to Ussher's calculation of his famous date. In 1616, Samuel Purchas employed this system for dating various events recorded in the Old Testament, such as the Noachian flood or the construction (and destruction) of Solomon's Temple.[31] In 1648, the Presbyterian physician (and defender of William Harvey's discovery of blood circulation), Roger Drake, carefully explained how to use the "famous and most artificial *Julian Period*" to "find out the year of the world."[32] In 1665, the Julian system received its scientific imprimatur when The Royal Society reported "a new and very easy Method" for calculating the year of the Julian Period, which had been invented by the French Jesuit mathematician Jacques de Billy.[33]

We might assume that this method of reckoning time has no place in the modern world. But this is very far from being the case. In the early days of digital computing, when computational memory was a fraction of what is available today on even a modern cell phone, using the Julian system enabled lengthy time periods to be calculated from the present until well into the future. So, in October 1957, the Smithsonian Astrophysical Laboratory used a modified Julian system to calculate the future orbit of the recently launched Soviet Sputnik satellite until August 7, 2576 CE, i.e., 619 years into the future.[34] The calculation soon proved to be redundant, since Sputnik survived for just three weeks before falling back into the Earth's atmosphere. But the fact that a time system that was originally devised by a sixteenth-century chronologer to compare ancient Greek and Roman history with biblical, Persian, Babylonian, Jewish, and Egyptian sources was used in the mid-twentieth century to calculate the predicted orbit of the first artificial Earth satellite is one of the unsung ironies of the history of both science and religion.

Far from the popular idea that Ussher was foolishly doing no more than adding up those lengthy lists of generational "begats" to be found in Genesis to calculate pre- and post-diluvial time, he was grappling with an immensely complex problem—that of trying to harmonize competing calendrical systems both with the scriptural text, historical astronomical data, and the records of classical antiquity. Ussher possessed "an erudition seldom matched by that of his critics."[35] To this end, his chronological project deployed "his vast knowledge of ancient languages, calendars, and history and match it to his biblical scholarship to construct a comprehensive chronology which linked together biblical and ancient history."[36] At the same time, Ussher's research was fired by the same dedication to "quantitative precision" that informed the work of late Renaissance astronomers and mathematicians, even when the data he was using was "fraught with problems of incomplete, ambiguous, or incompatible records."[37] So, his work was comparable to that undertaken, for example, by the German astronomer, Johannes Kepler, who devoted nearly as much time to computing chronology as he did to calculating the orbits of the planets, and whose *Tabulae Rudolphinae* (Rudolphine Tables) of 1627 were equipped not only with the means of recording the positions of heavenly bodies, but also chronological tables and the results of his work on ancient calendars and dating systems.[38]

Ussher's chronology was devised to "provide a biblical schema that serves Protestant providentialism."[39] But the archbishop was nevertheless scrupulous in recording only what could be verified from his sources and his calculations. This, in essence, was the proto-scientific method advocated by the nascent Royal Society. Hence, Ussher's use of the Julian system for calculating time, for (as he wrote):

> in divers Ages and Nations, diverse Epochs of time were used, and several forms of years: Here it's necessary that some common and known account should be observed, to which the diversity of the rest may most appositely be reduced. And to us there is no measure of time more known, and more accommodatious to the common collation of times than the form of the Julian Years and Months.[40]

The phrase "diverse Epochs of time" referred not just to different calendrical systems, but also to alternative divisions of time: the four-year cycles of the ancient Greek Olympiad; the constituent (and declining) "ages" of human existence (golden, silver, bronze, and iron) inherited by the Renaissance from Greek and Roman mythology; or St. Augustine's Six Ages of the World.[41] Ussher recognized, too, that units of time were not stable across different cultures. It was the deployment of a "common collation of times" which explains Ussher's otherwise puzzling calculation that, although the creation of the world (and of time itself) began in Year 1 of *Anno mundi* (4004 BCE), this corresponded to the 710th year of the Julian period. The Julian period was a purely synthetic timescale—a means of measuring an abstract idea of elapsed time rather than human history, which was (or should be) a procession of dateable recorded events. Hence, Ussher's 709 Julian years prior to the creation of time and history can be understood as virtual time—time which did not yet exist, but which, according to the Julian system, could nevertheless be computed.

Ussher's commitment to reason, to the comparison of his sources (which included astronomical tables and records of eclipses), and to his calendrical calculations could lead him to conclusions which (to the modern mind) might seem astonishing. Indeed, they seemed, on occasion, even to have astonished the archbishop himself. How long, for example, did Eve and Adam linger in paradise before they were expelled from the Garden of Eden? In John Milton's *Paradise Lost* (1667), the prelapsarian idyll seems, essentially, timeless, since death had not yet been introduced into the world. But Milton, like Ussher, was also committed to an idea of sacred chronology. The events recounted in *Paradise Lost*, a poem which begins, of course, *in medias res*, occupy the space of just over one calendrical month, beginning with the first exaltation of Christ, recounted in Book V of the poem. The creation of Eve and Adam takes place on Day 19 of the poem (Book VII), the Fall on Day 32 (Book IX), and the expulsion from Eden (and with it the conclusion of the poem) on Day 33 (Book XII).[42] Thus, according to Milton, Adam and Eve sojourned in paradise for just fourteen days.

But Ussher's investigations led him to a rather different conclusion. Having recounted how Eve had been "beguiled by the Serpent, and the man seduced by the woman," Ussher reported the results of his calculations: "It is very probable, that *Adam* was turned out of Paradise the selfsame day that he was brought into it, which seemeth to have been upon the tenth day of the world (answering to our first day of *November*, according to [the] supposition of the Julian Period)."[43] Rather than Eve and Adam enjoying a luxurious Miltonic fortnight in Eden, Ussher concluded that our primal parents barely spent a single day in paradise. But note the speculative tone which Ussher adopts as he leads his reader to this conclusion: "it is very *probable* [...] which *seemeth* to have been" (my emphasis). The calculation, moreover, is based on a "supposition" (i.e., a theory), which may or may not prove to be true. Far from appearing as a dogmatic, literal-minded, fundamentalist, Ussher was anticipating the probabilistic language of later seventeenth-century natural philosophers. This was an approach which Thomas Sprat had commended in his *History of the Royal Society*—a work published in the same year (1667) as *Paradise Lost*—writing that it was the task of natural philosophy to proceed with "circumspection and modesty" and to be "cautious to shun the overweening *dogmatizing* on causes" as well as to avoid "*speculative scepticism*," but, rather, to promote "just deliberation."[44]

The Problem with Large Numbers

It has been claimed that, prior to the publication of *La Scienza Nuova* (The New Science, 1725) of Giambattista Vico, "there was little concept of the passage of time in the study either of human society or Nature."[45] But, as the case of either Scaliger or Ussher indicates, this seems unlikely. The 5,600 or so years, which was the age of the Earth according to Ussher's calculations when they were first published in 1650, can also be expressed as fifty-six centuries. The 100-year unit of the century was, however, a relatively novel approach to organizing and dividing time, having been introduced by Protestant historians—"Centuriators" as they became known—at Magdeburg in 1559 when the multi-volume *Historia Ecclesiae Christi* (History of the Christian Church) began to appear, in which each hundred-year cycle was allotted a single volume. Despite this new unit of time being devised as "a technical stopgap" in a multi-volume work which set out to show how the Catholic Church had falsified the Gospels, it "soon made historians more inclined to arrange events and documentary evidence in chronological order, even when [...] no firm date was known."[46]

Ussher's calculation of a juvenescent Earth proved to be not merely incorrect but wrong by a factor of nearly 757,000 (i.e., the Earth is now held to be nearly 757,000 times older than the 6,000 years to which Ussher's calculation amounts today). Why was it not possible for Ussher and his contemporaries to sense something of the enormous depth of time which measured the actual history of the Earth? The usual (perfectly sensible) answer to this deliberately naïve question is that chronologists such as Ussher or Scaliger were relying entirely on written records of human history, which covered (as we now know) just a minute fraction of the Earth's natural history. Not until the creation of an "alternative numerical chronology" in the second half of the eighteenth century did the Earth acquire its own, independent, history: "a new and vastly extended timescale, anchored in the rock strata and fossils of the Earth's crust"—a timescale to which the Kentucky Creation Museum is implacably opposed.[47]

But perhaps there is another, rather more nebulous, gulf that separates our world from that of the chronologists and that helps us to better understand why the figure of 4004 BCE seemed perfectly plausible in the mid-seventeenth century. Such a gulf has nothing to do with as yet unknown geological data, nor with a literalist approach to scripture, but with the measurement of time itself, together with the conceptual difficulty of grasping the reality of very large numbers. In the sixteenth century, the mathematical sciences were "ranked low in the hierarchy of the academic disciplines when compared to theology and philosophy."[48] This is not to deny the sophistication of medieval and Renaissance mathematicians, who were very much aware of the Greek and Arabic mathematical traditions.[49] But mathematics was primarily understood as a practical discipline, useful for inculcating "mechanical" skills in navigation or commerce. Hence, there was very little wider understanding of the abstract calculation of large numbers prior to the later seventeenth century, when "mathematical modes of explanation and mathematization" began to emerge as "potential instruments of power and a source of valuable technical and socio-political knowledge."[50]

Ussher's Julian timescale gave him (as we have seen) a "block" of time with which to operate—the 7,980 years of the Julian period. To the modern geologist or cosmologist, used to calculating time in units of millions or billions of years before the present (BP), a timescale encompassing just 7,980 years is a risibly inadequate tool.[51] But this is to ignore the ways in which early modern people not only conceived of the flow of time on a different and much more compressed scale but also how they approached the entire problem of handling very large numbers. Thus, when Drake published his *Sacred Chronology* in 1648, the space of 4,000 years measuring (as he believed) the span of time from the creation of the world to the "Passion of our Blessed Saviour" was understood to be a "*Vast Body of Time.*"[52] Burton, perched uncomfortably between the "two spires of the cathedral of English scientific thought" (i.e., Francis Bacon and Thomas Hobbes), as he has been described, helps us to appreciate this problem more clearly.[53] In his *Anatomy of Melancholy*, Burton loved to perform the most abstruse and seemingly pointless calculations, such as (for example) the fact that "ten words may be varied 40,320 several ways," or investigating "how many men may stand one by another in the whole superficies of the Earth," when each is assigned a single square foot: the answer being 148,456,800,000,000—a calculation which also expressed the surface area of the Earth not in square miles, but in the superfluous units of square feet (2:92).

But nothing engaged Burton's mathematical imagination more fully than calculations associated with the movement and dimensions of the heavens. The movement of planetary bodies, for example, was driven by an "incomprehensible celerity in 24 hours" so that "every point of the Firmament and in the Æquator must needs move 176,660 [miles] in one 246th part of an hour." This speed, he maintained, was equal to that obtained by an arrow, shot from a bow, and compassing the Earth seven times in the space of time that is needed "to say an *Ave Maria*"—a time interval that was reckoned to amount to around fifteen seconds. Or, again, the enormous distance travelled by the rotating "Firmament" (by which Burton meant the eighth sphere of the fixed stars in a geocentric model of the universe) was such that a man "could not ride so much ground going 40 miles a day in 2,904 years, as the said Firmament goes in 24 hours" (2:50). How far is the distance between the Firmament and the Earth's surface? Burton performs the calculation for us: "if a stone could fall from the starry heaven, or eighth sphere, and should pass every hour an hundred miles, it would be 65 years, or more, before it would come to ground, by reason of the great distance of heaven from Earth, which contains as some say 170 millions 800 miles" (1:183). What is the distance between the "sphere" of the planet Saturn and the "Firmament?" The answer is a yawning emptiness: "an incredible and vast space or distance," amounting to "7,000,000 semidiameters of the Earth [...] void of stares" (2:51).[54] Other calculations, which might seem to us to hover on the brink of the absurd, were quite possibly informed by Calvinist ideas of predestination. What is the ultimate capacity of hell for the damned? Burton's answer to this infernal mathematical problem is "a sphere able to hold eight hundred thousand millions of damned bodies (allowing each body six foot square) which will abundantly suffice," since he was certain that there would be no more than "*centies mille milliones damnadorum*" or 100,000,000,000 souls to be damned (2:41).

Burton's extraordinary mathematical calculations are designed to dazzle the reader by evoking the cognitive impossibility of conceiving of such immense speeds, or distances, or spaces on any kind of human scale. In this respect, they seem to anticipate the Kantian "mathematical sublime"—a sense of enormity which transcends the limits of the human imagination.[55] Such figures (as Burton puts it in the *Anatomy*) are "*supra humanam cogitationem*, beyond humane conceit" (2:50). His sub-calculations (as we might term them), in which vast figures are wreathed in digressive secondary calculations, emphasize their incomprehensibility: an arrow shot from a bow and impossibly circumscribing the Earth, again and again, in a few seconds; a man riding on horseback for nearly 3,000 years; or a stone dropping through the void for almost the span of a human life.

Burton was accounted by his contemporaries "a learned mathematician."[56] But he also understood that such numbers presented almost insuperable difficulties of comprehension. Those difficulties allow us to place Ussher's 5,600-year lifespan of the world within a broader cultural context. So, to Ussher and his contemporaries, this calculation probably expressed a much longer duration of time than it does to us. This is not because the numerical value associated with the word "thousand" has shifted but rather because very few sixteenth- and seventeenth-century people ever had the occasion to count in thousands or hundreds of thousands, let alone millions, with the exception of some highly specialized groups, such as merchants, treasurers, accountants, and printers. Early modern printers, for example, were one of the few groups of artisans who had to understand how to count in the scores of thousands, when they estimated the number of sheets of paper that might be required to produce multiple copies of a book.[57]

When it came to calculating very large numbers, most early modern people were likely to have been in a similar condition to the "*Americans*" (i.e., Indigenous peoples of South America) with whom John Locke claimed to have discussed calculation—an encounter he recorded in his *An Essay Concerning Human Understanding* (1689) in the chapter entitled "Of Number." Locke's "Americans" were people of "quick and rational parts," but they could not "by any means count to 1000 nor had any distinct Idea of that number, although they could reckon very well to 20." Leading "a needy simple life, unacquainted either with Trade or Mathematics," they possessed no words "to stand for a thousand" so that, when they "discoursed" of greater numbers, they would "shew the Hairs of their Head, to express a great multitude which they could not number."[58]

But in this respect, Locke argued, his "*Americans*" were not much different from the condition of Europeans when confronted by very large numbers.[59] However, large numbers represented a terminological rather than a mathematical problem. "And I doubt not," Locke wrote, "but we ourselves might distinctly number in Words a great deal farther than we usually do, would we find out some fit denominations to signify them by." Citing an impossibly vast, confected, number, which stretched from "Unites" via "Millions," "Billions," "Trillions" all the way to "Septillions," "Octillions," and "Nomillions," Locke described how "the ordinary way of naming this Number in *English* will be the often repeating of Millions, of Millions, of Millions, of Millions [...] In which way, it will be very hard to have any distinguishing Notions of this Number" (208–9). The problem, it seemed, resided in the human imagination, which was unable to grasp conceptually the "Idea" of the number.

And the same problem persisted in the calculation of time. For, as Locke pointed out, the understanding of the passage of time depended on "different Ideas of Duration which are settled in our Minds" (326). Contemplating the question of what constituted time, Locke was tempted to echo St. Augustine: "*si non rogas intelligo* (which amounts to this; the more I set myself to think of it, the less I understand it)" (181). The division and measurement of time was, for Locke, an arbitrary process, so that he suggested that "the freezing of Water, or the blowing [i.e., flowering] of a Plant, returning at equidistant Periods in all parts of the Earth, would as well serve Men to reckon their Years by as the Motions of the Sun" (188). If the accepted division and measurement of time was subject to revision, then so, too, was the calculation of the world's origin. Believing that the "Duration of the World from its first existence to this present year 1689 to have been 5639 years," Locke was, however, willing to concede that there might be other calculations which might be equally true: the 23,000 years of the "Ægyptians of old" or even the "3,269,000 years" of the Chinese (194). The core of the problem was not the accuracy of the calculation but "the measuring the duration of anything by Time" and this depended on "the Idea of the length of any regular periodical Appearances that which we can in our Minds apply to Duration" (195).

Millions, Billions, and Trillions

Modern scientists, when they record the measurement of either the very small or the very large, use scientific notation. For example, an electron's mass has been calculated as 0.000 000 000 000 000 000 000 000 000 000 910 938 356 kg, which, in scientific notation, can be expressed as $9.10938356 \times 10^{-31}$ kg. Although Archimedes is usually credited with the invention of an early form of scientific notation, when (in the third century BCE) he calculated that the number of grains of sand required to fill the "universe" was (as we would now express it) 1×10^{63} (i.e., one followed by 63 zeros), scientific notation only entered common use in the mid-nineteenth century.[60] For all the apparent precision and conciseness of this method of recording numeration, most of us non-scientists are perhaps not much better than our early-modern forebears when it comes to understanding the "Idea" (to echo Locke) of very large (or very small) numbers.

Thus, when we are told that the Earth was formed around 4.54 billion years ago, we know that this seems to be a very large quantity of elapsed time on any human scale. But it is only when we begin to translate this figure into a different or parallel time scheme that we begin to grasp just how immense a passage of time is being measured, which was Locke's point about "different ideas of Duration." The figure 4.54 billion, using the now standard American short form of a billion (i.e., one thousand million), can be expressed as 4,540,000,000, or 4.54 thousand million, or (in scientific notation) 4.54×10^{9}. If we were to use the unit of the century, then the Earth's age can be reckoned as just under 45 ½ million centuries. But even to the modern mind, these figures do not make much sense, any more than does the idea of a light year with which astronomical distances are calculated today, even if we habitually use the term in every day speech to express (in the words of the *Oxford English Dictionary*) "an indefinitely great distance" or "an indefinitely long time."[61] How long a period of time, for example, does a billion seconds measure? Ten years? One hundred years? One thousand years? Ten thousand years? The correct answer is 31,687 years, 17 days, 22 hours, 45 minutes, and 5 seconds, according to the British scientist and inventor Sir Henry Bessemer, who, in a letter to *The Times* ("A Billion Dissected," January 28, 1878), announced this calculation to demonstrate that "it is no easy matter to bring under the cognizance of the human eye a billion objects of any kind." Commenting that "a billion is a fearful thing" and observing that "few can appreciate its real value," Bessemer concluded that as for trillions and quadrillions (and perhaps echoing Shakespeare's Hamlet): "they are simply words: mere words wholly incapable of impressing themselves on the human intellect."[62]

If it is difficult for us, today, to understand the true dimensions of a billion, even when we read that (say) the net worth of the world's richest individuals can be calculated in terms of tens or even hundreds of billions of dollars, then how much more difficult was it for early modern people to understand very large numbers?[63] This problem was addressed in the many arithmetic textbooks that were published throughout the sixteenth and seventeenth centuries that helped to introduce principles of calculation to the "unlearned" (i.e., those lacking Latin). But the ways in which those principles were explained also hints at the difficulties associated with counting large figures. In Thomas Willsford's popular *Arithmetic, Natural and Artificial* (1656), for example, readers were taught "numeration" to demonstrate

> the value or quantity of any number given; or how to express in proper figures any number assigned or propounded and to declare the quantity of each figure, according to its place or position, their contents increasing by degrees ascending from the right hand towards the left, as do the letters and characters of the Hebrews, and all the Eastern people, from whence this Art of Number extracts its original.

The accompanying tables demonstrated how to write very large numbers according to their "degrees" or quantities:

> In this amplification of Numbers you must observe the figure on the right hand denotes simply itself, as 2 two; 4 four; 6 six; 9 nine; 0 nothing &c. the next degree, or second figure towards the left hand is ten times its own quantity; the third place a hundred times; the fourth degree a thousand times; the fifth place ten thousand times; the sixth figure in order is a hundred thousand times its own number; the seventh place a million; the eighth ten millions; the ninth place a hundred millions;

> and so proceeding without end, increasing every degree in this manner following, reiterating Unites, Tens, and Hundreds.[64]

Willsford's primer (as we might now term it) was not aimed at children but at literate adults such as "merchants, citizens, sea-men, accomptants, &c" who needed to develop practical mathematical skills.[65] But, as his careful description of how to inscribe larger and larger figures might suggest, Willsford knew that this was probably unfamiliar terrain for his readership.

Although the words (and hence the arithmetical concepts) of billions and even trillions clearly existed in the early modern world, people seem to have had very little use for them. The *OED* tells us that the word "billion" first appears in print in the English language in 1611, where it was defined as a "million of millions," while "trillion" first appears in 1635.[66] But the actual usage of these terms in the seventeenth century is vanishingly slight. Burton, who, as we have seen, delighted in calculating very large numbers, frequently had recourse to the idea of a "million" or "millions" in the *Anatomy*, but the terms "billion" and "trillion" (for all that he might have found them useful) seemed not to have formed any part of his word stock. The commonplace "million" appears occasionally in earlier English sixteenth-century printed texts, but it is not until the mid-seventeenth century that the word was widely used. And even when the words "billions" or "trillions" do make an appearance, they seem to have been deployed not as a means of calculation, but rather as a way of expressing what is (figuratively speaking) inexpressible. So, in 1695, the cryptographer, inventor, and engineer, Sir Samuel Morland, set about calculating not the number of angels who could (proverbially) dance on the head of a pin, but exactly how many angels God had created on the first of the six days of creation. To us, the calculation may seem (once again) absurd, but it demonstrates the conjunction between a rational science of investigation based on measurement or quantification, and the revealed truth of scripture which Ussher, too, was pursuing. Combing the Bible for passages which mentioned the size of the angelic host ("*More than* 12 *Legions* […] A Thousand Thousands ministered unto him, and Ten Thousand times Ten Thousand stood before him"), Morland eventually concluded that the calculation could not be performed:

> What Number should we think a fit Retinue for the KING of KINGS, and LORD of LORDS? *Myriades Myriadum,* will be too small a Sum: And so will our ordinary Terms of Arithmetic, and we must be forced to borrow some of the learned Commentator upon *Archimedes* his *Arenarius,* and say, Billions, *Trillions*, Quadrillions, &c. of blessed Angels to attend his Throne, and to make up the Court of Heaven.[67]

Morland's billions, trillions, and quadrillions are designed to express what cannot be calculated using the "ordinary terms of Arithmetic." His "*Myriades Myriadum*" (i.e., 10,000 × 10,000 = 100,000,000) is not a real number but rather an expression of the uncountable since, in the seventeenth century, the English "myriads" was a synonym for "countless."[68] In his cryptographic work, Morland liked to deploy vast numbers to evoke what could not be counted, observing (for example) that "the *Permutations* of 24 Letters are so numerous, that a Thousand Millions of able Clerks, in a Thousand Millions of Years" would not be able to transcribe them, so that one code which he had devised would require "*Quadrillions* of *Trillions,* of *Billions* of *Methusalem's* years" to crack without the necessary key.[69]

In everyday use, early modern people tended to express larger numbers as multiples of a thousand, using indefinite intensifiers such as "many" or "tens" or "scores." This, indeed, was Ussher's preferred method of counting. In this, they were following the example set by the Bible. Only once in the entire King James Bible does the word "million" occur, when Rebekah, the wife of Isaac, learns that she will become "the mother of thousands of millions" (Gen. 24:60), which is equivalent to saying, once more, "incalculable."[70] The largest concrete figure mentioned in the Bible is two million, but it is transcribed as "two hundred thousand thousand"—the number of "the army of the horsemen" let loose at the blowing of the sixth trumpet heralding the apocalypse (Rev. 9:16). William Shakespeare uses the term "million" in his plays sparingly compared to his frequent recourse to thousands and multiples of thousands. And when he does deploy the word, it tends to express an infinite number rather than a realizable figure, as in Robin's exclamation that for every one individual "holding troth, / A million fail" (*Midsummer Night's Dream*, 3.2.93-4), or else it denotes simply a numberless multitude, as in Hamlet's observation that a play he had once watched "pleased not the million" (*Hamlet*, 2.2.439).[71] For, as Paula Blank has argued, "numbers provided Shakespeare with a system of signs in which the meaning of a 'figure' depends on its comparison with others" to register "values that are, and can only be, relational."[72]

Writing in the late 1650s and early 1660s, Milton frequently evoked vast numbers when describing the angelic armies who clash in the pages of *Paradise Lost*: the "millions of flaming swords" drawn to greet Satan's speech recommending war in Book I of the poem, or the "armed saints" of heaven, led by Michael and Gabriel, whom God numbers by "thousands and by millions ranged" in Book VI (I. 664; VI. 48). However, just as the Bible preferred a numbering system using thousands rather than millions, so Milton, just like Ussher, much preferred to evoke angelic hosts using multiples of a thousand: the "ten thousand thousand [i.e., ten million] ensigns high advanced" born by the heavenly army in Book V of *Paradise Lost*, or the "ten thousand thousand saints" who accompany Messiah in Book VI (V. 588; VI. 767).[73] For Milton, a thousand, it seems, was more imaginable, tangible even, than a million, while numbering things (satanic armies, angelic hosts) in multiples of a thousand, paradoxically, seems to give a greater sense of the size of vast crowds, rather than using the unimaginable "million."[74] Indeed, Milton tended to use the terms "innumerable" or "innumerous" when it came to describing enormous quantities of anything: stars, fallen angels, or the future population of the human race.[75]

When early modern people who were not mathematicians or merchants (or printers) tried to count a very large number of objects in a series, they tended to wildly overestimate the result, suggesting that their sense of numeration in proportion to a given space or area was hazy. In the early summer of 1643, for example, the Scottish traveler William Lithgow undertook a twelve-hour "perambulation" of London's parliamentarian defenses that were built in anticipation of a threatened royalist siege in the first months of the Civil War. These defenses, Lithgow calculated, were designed to protect a city which, he believed, consisted of 500,000 dwellings, housing a population of "three millions of souls."[76] In fact, this was probably seven times the actual population of London in the 1640s, which has been estimated to have amounted to around 400,000 people.[77] This was a problem which would remain largely unresolved until the advent of "political arithmetic" pursued in John Graunt's *Natural and Political Observations Made upon the Bills of Mortality* (1662), which offered a semi-reliable means of calculating human populations and necessarily involved the statistical manipulation of large figures with a degree of what we would think of as accuracy.[78] But even after Graunt's *Observations* were available, it was still the case that estimates of populations tended to be overcounted by very large factors. One late seventeenth-century calculation of the population of the London parish of St. Giles in the Fields, for example, reputedly the most populous in England, gave a figure of "1,000,000 souls besides strangers who are very numerous" which was an impossibility.[79]

Eternity

The general early modern preference for calculating large numbers using thousands rather than millions, let alone billions, helps us to appreciate the chronologers' sense of (by today's standards) foreshortened time. Ussher's date of 4004 BCE for the creation of the universe was the product of a world in which quantifying *anything*, not just time, on a scale involving millions rather than thousands, was extremely unusual. As for billions, although the word existed, the "Idea," as Locke concluded, was almost unimaginable. And yet, the mid-seventeenth century is also the moment when the lemniscate, or infinity symbol (∞), possibly based on the Roman numeral for 100 million, was first introduced into mathematical notation. In his *De sectionibus conicis* (Of Conic Sections, 1656), the English mathematician John Wallis first used the lemniscate when he wrote that any plane could be considered as consisting of an infinite number of parallel lines.[80] Locke, too, puzzled over the ideas of both eternity and infinity as "an endless addition of number" in his *Essay Concerning Human Understanding* (212).[81] But for Locke, "the actual Idea of an infinite number" was an "absurdity." Whether considering infinite space or endless duration, the only recourse was "the repetitions of Numbers, as of millions of Miles, or Years […] wherein the Mind loses itself" (214–15).

These, of course, were understood as theological as much as they were mathematical questions. In *The City of God*, St. Augustine had pondered the same question that would later trouble Burton: "But why did the eternal God decide to make heaven and earth at that particular time, and not before?" Perhaps, as Augustine grudgingly conceded, creation might be considered "an eternal process" rather than accepting the theologically unacceptable alternative that God suddenly conceived of the idea of creating the world as a "random, fortuitous act."[82] The idea that the world might be immensely old, or even eternal, would not have surprised medieval and early modern theologians or philosophers, who conjured with the notion of an "ancient Earth of indefinite age" whether in the Christian or the Muslim traditions.[83] In this respect, the chronologists were contesting a view of the duration of the Earth's history: that it might be (as we have seen) unknowable, or that it might be extremely ancient, or even that it might be sempiternal (existing indefinitely within time) or eternal (existing atemporally, that is, beyond the constraints of time altogether).[84]

But there were other ways of imagining either eternal or sempiternal duration. "I saw Eternity the other night / Like a great *Ring* of pure and endless light." So, Henry Vaughan began his poem "The World," first published in the same year (1650) that Ussher announced his calculation of the world's beginning. Vaughan's "eternity" operates like an unending, luminescent, armillary sphere—the ancient, mechanical model of celestial motion and one of the most complex devices known to Renaissance astronomy. Beneath this "*Ring* of pure and endless light," Vaughan's poem continues:

> Time in hours, days, years
>   Driv'n by the spheares
> Like a vast shadow mov'd, In which the world
> And all her train were hurl'd.[85]

In Vaughan's poetic vision of the universe, time, measured in the passage of "hours, days, years," is an abstract "vast shadow" cast by the circular light of eternity, into which the world and all that it contains is "hurled" in what seems like a foretaste of the "wild abyss" of Chaos in Milton's *Paradise Lost*—a space without measurable "dimension" in which "length, breadth, and highth, / And time and place are lost" (II. 910, 892–4).

Vaughan's unforgettable vision of time and eternity was paralleled in Thomas Traherne's rapt, mystical poetry, which abounds in visions of the infinite and the eternal.[86] In (for example) "The Salutation"—a poem probably composed in the 1660s—we encounter a sense of deep but mysterious time, as Traherne imagines his own created, infant, soul, patiently waiting, silently, "So many thousand thousand years" (i.e., a million years) for embodiment, as he lingers "a nothing from eternity."[87] The measurement of time and the gaping abyss of eternity figures, too, in what is (today) Andrew Marvell's most familiar poem: "To His Coy Mistress," probably composed in the late 1640s or early 1650s, but not published until 1681. In Marvell's poem, the speaker assures the "coy mistress" of the poem's title that he would love her "ten years before the flood" and that she might refuse his blandishments "till the conversion of the Jews," which is to say that the poem's imaginative chronology stretches from the antediluvian past into something like the present, when some held that the readmittance of the Jews into England under the Cromwellian regime heralded the coming Millennium—the return of the Messiah. Promising to devote lengthy stretches of time in adoring the mistress's eyes and forehead (100 years) and breasts (200 years for each), the poem launches into a sexually knowing evocation of a comically lengthy period of time—thirty thousand years—in admiring "the rest." But the future is a bleak and unknowable desert "of vast eternity" in which the would-be lovers will themselves be slowly devoured by the "slow-chapped power" of Time. Marvell's poem wittily plays with the idea of vast stretches of computable time slowly passing, but also with compressed images of time speeded up, since (as the poem concludes) "though we cannot make our sun / Stand still," as had Joshua in the Old Testament, "yet we will make him run."[88]

Conclusion: Changelessness

But what if God measured the passage of time in ways that were unfathomable to human reason? For, as the Psalmist had proclaimed, "a thousand years in thy sight are but as yesterday when it is past" (Ps. 90:4), which was echoed in the New Testament admonition: "be not ignorant of this one thing, that one day is with the Lord as a thousand years, and a thousand years as one day" (2 Pet. 3:8). It might be thought that these scriptural texts offer the Kentucky Creation Museum a means of recalibrating its conception of time altogether. If we follow the literal Petrine formula of one day "with the Lord" being equivalent to a thousand years, then Ussher's date for creation could be pushed back by nearly 1.5 billion (solar) years. But, true to its commitment to a Young Earth ideology, that temptation is firmly resisted by the Creation Museum. According to the "Statement of Faith" to which all employees of AiG must subscribe: "The days in Genesis [...] are six consecutive, 24-hour days of creation; the first day began in Genesis 1:1, and the seventh day, which was also a normal 24-hour day, ended in Genesis 2:3."[89]

Together with its more recent offshoot, Ark Encounter, also situated in northern Kentucky, which offers a full-size replica of Noah's Ark for visitors to explore, AiG has commandeered the so-called conflict thesis, promulgated in the late nineteenth century. That thesis postulated the "unavoidable hostility and incompatibility" between science and religion, ignoring almost entirely the "complexity of the historical engagement" between the two, to promote a "binary view of premodern culture stubbornly clinging to the letter of the Bible, as opposed to the later triumph of a scientific world view."[90] Far from clinging "stubbornly" to the primacy of the biblical record of the past, as we have seen, Archbishop Ussher struggled to reconcile the scriptures with the products of human reason. In fact, for all that Young Earth Creationists embrace Ussher's seventeenth-century chronology, this modern doctrine bears hardly any resemblance whatsoever to the type of quantitative approach to understanding the flow of time which was pursued by the chronologists. Indeed, what was radical, and entirely consistent with the emergent "New Philosophy" of the seventeenth century, were the efforts of chronologists such as Scaliger and Ussher to introduce an idea of measurable—that is, quantifiable—time into the Bible's account of creation and to make it consonant with human history. Conversely, anchored to a literalist version of scripture, Young Earth Creationism not only relies upon what is now understood to be a drastically foreshortened time scale within which to understand the origin of life but also on the primacy of (highly selective) eyewitness accounts to record events either in the past or the present—a procedure that it terms "observational science."[91] "Ultimately," it explains, "the best dating method is a trustworthy eyewitness, and the Bible provides the testimony of the only perfect eyewitness, our Creator God."[92]

Ironically, this skeptical view of deductive reasoning seems to echo Milton's verses in *Paradise Lost* with which this essay is headed, with its rhetorical question, posed by Satan: "Who saw / When this creation was?" (V. 856–7). The answer is, of course, nobody, since nobody, other than

God, could be witness to the moment of creation, for all that both Adam and Eve, in Milton's poem, recall their first moments of consciousness *after* their emergence into being. Or, to express this view in the words of the AiG "Statement of Faith:"

> The account of origins presented in Genesis 1–11 is a simple but factual presentation of actual events, and therefore, provides a reliable framework for scientific research into the question of the origin and history of life, mankind, the Earth, and the universe […]. No apparent, perceived, or claimed evidence in any field of study, including science, history, and chronology, can be valid if it contradicts the clear teaching of Scripture obtained by historical-grammatical interpretation.[93]

This uncompromising statement of belief would probably have appalled the seventeenth-century Archbishop of Armagh, who spent a good portion of his life struggling to reconcile the scriptures with the countervailing evidence provided by his extra-biblical sources, including natural phenomena, which he recognized as offering authentic records of the past.

Young Earth Creationists are hostile to any form of evidence other than that based on their interpretation of (by their lights) a single, authoritative text, dismissing the products of modern science as godless delusions, since fallen human reason is inherently fallible. So, the "naturalistic worldview" of modern "evolutionists" is also held to be endowed with "utter hopelessness and meaninglessness" compared to "the true hope and meaning revealed in God's word."[94] Milton's Satan believes in the self-willed and self-authorized timelessness of his own unchanging existence. Changelessness, too, lies at the core of Young Earth Creationism: "The various original life forms (kinds), including mankind, were made by direct, supernatural, [and] creative acts of God (i.e., not by natural, physical processes over millions of years) […]. Only limited biological changes […] have occurred naturally within each kind since creation (i.e., one kind does not change over time into a different kind)."[95] By the same token, in proclaiming that he can know "no time when we were not as now; / Know none before us" (V. 859–60), Milton's Satan seems to stand shoulder to shoulder with the Kentucky Creationists. Both he and they are convinced of their essential changelessness, as one wages war against God, and the other against modernity.[96] The outcome of the former conflict, according to Milton, was foreordained. As to the outcome of the latter, at least in the contemporary United States, at a moment when science, scientific institutions, and even scientific reasoning are being subjected to an ever-more strident assault, only time itself will tell.

Notes

I would like to thank Professor Valerie Allen of the Department of English, John Jay College of Criminal Justice CUNY, together with Professor Anneke Bart, Chair of the Department of Mathematics and Statistics at Saint Louis University, and Professor Derek Hirst of the Department of History, Washington University in St. Louis, for their help in preparing this chapter.

1 Gretchen Jennings, "Creationist 'Museums' Are Not Museums," 73.

2 Susan L. Trollinger and William Vance Trollinger, Jr., *Righting America at the Creation Museum*, 13.

3 Anon., *Journey Through the Creation Museum*, 31. On radiometric dating, see von Gunten, "Radioactivity: A Tool to Explore the Past." On current estimates of the age of the Earth (4.54 ± 0.05 billion years) and the history of scientific dating of the Earth's origin see, *The Age of the Earth: from 4004 BC to AD 2002*, ed. Lewis and Knell.

4 These impressions are based on a visit to the Creation Museum, undertaken by the author in March 2025.

5 Creation Museum Website, https://creationmuseum.org/; Trollinger and Trollinger, *Righting America*, 13.

6 Trollinger and Trollinger, *Righting America*, 200–1. On Adams's *Synchronological Chart*, see Daniel Rosenberg and Anthony Grafton, *Cartographies of Time*, 173.

7 Answers in Genesis, "The Annals of the World, Written by James Ussher," accessed December 2, 2024. https://answersingenesis.org/store/product/annals-world/?sku=10-3-107.

8 Colenso, *The Pentateuch and the Book of Joshua Critically Examined*, vii–viii. On the Green/Colenso controversy, see Ronald L. Numbers, "'The Most Important Biblical Discovery of Our Time': William Henry Green and the Demise of Ussher's Chronology."

9 Green, *The Pentateuch Vindicated from the Aspersions of Bishop Colenso*, 128.

10 Quarles, *An Elegy on the Most Reverend & Learned James Ussher;* Anon., *An Elegy on the Miraculously Learned and Much-Lamented Bishop of Armagh*.

11 Bernard, *The Life & Death of the Most Reverend and Learned Father of Our Church Dr. James Ussher*.

12 Parr, *The Life of the Most Reverend Father in God, James Ussher, Late Lord Archbishop of Armagh*, sig. K1r.

13 Gordon Campbell, *Bible*, 121.

14 Ussher's chronological calculations were not, however, reproduced in American imprints of the King James Bible until 1917, when the second edition of the "Scofield Reference Bible" appeared, which embedded Ussher's chronology in the American "popular evangelical imagination." See Campbell, *Bible*, 245.

15 See W. W. Wroth and Philip Carter, "Hales, William."

16 Green, "Bases of Scripture Chronology," 97.

17 James Barr, "Why the World Was Created in 4004 B.C.: Archbishop Ussher and Biblical Chronology," 581.
18 Gould, *Eight Little Piggies*, 184–85.
19 Ussher, *Annales veteris testamenti*, sig. B1r.
20 Ussher, *The Annals of the World Deduced from the Origin of Time*, sig. B1r.
21 Grafton, "Joseph Scaliger and Historical Chronology," 169–70. On Scaliger's development of the Julian scheme, see Arno Borst, *The Ordering of Time*, 104–6.
22 Martin J. S. Rudwick, *Earth's Deep History*, 12.
23 Du Bartas, *The Divine Weekes and Workes*, ed. Snyder, I. 112.
24 Raleigh, *The History of the World*, sig. B2r. A *Targum* is the (originally oral) Aramaic translation of the Hebrew Bible, while *Sapientia* refers to the idea of divine wisdom, connected to the concept of Logos or the Word of God.
25 Burton, *The Anatomy of Melancholy*, ed. Faulkner et al., 2:57–8. Hereafter cited parenthetically in the text by volume and page number.
26 Browne, *Works*, ed. Keynes, 2:399.
27 Ibid., 2:404, 410.
28 The 19-year lunar cycle, also known as the Metonic cycle or *enneadecaeteris* (from the Greek, ἐννεακαίδεκα, meaning nineteen), measures the period between which the monthly lunar phases recur at the same time of year. Reputedly first calculated by the mathematician and astronomer Meton of Athens in the fourth century BCE, this cycle was the basis of the Greek and Hebrew calendars. The 28-year solar cycle (the basis of the ancient Egyptian calendar) measures the period between which the distribution of the days of the week in each calendar month repeats. The fiscal indiction cycle was introduced as a dating system throughout the Roman Empire by Constantine in 312 CE and was adopted as a general Christianized dating system at the First Council of Nicea in 325 CE. See Grafton, "Joseph Scaliger and Historical Chronology," 181–4; Alexander Phillip, *The Calendar: Its History, Structure and Improvement*, 43–6; Robert Hannah, *Time in Antiquity*, 1–42.
29 In calculating this Julian Day Number, I have used the Julian Day Calculator at https://www.tondering.dk/claus/cal/julperiod.php. The calculator explains the basis of the calculation, using the target date January 1, 2000 CE, as follows: "From 4713 BC to AD 2000 there are 6712 years. In the Julian calendar, years have 365.25 days, so 6712 years correspond to 6712 × 365.25 = 2,451,558 days. Subtract from this the 13 days that the Gregorian calendar is ahead of the Julian calendar, and you get 2,451,545." For the purposes of calculating decimal fractions of a Julian "day," each "day" is held to begin at midnight, unlike astronomical "days" which begin at noon.
30 See Marcus Kuhn, "A Summary of the International Standard Date and Time Notation" at https://fits.gsfc.nasa.gov/iso-time.html.
31 Purchas, *Purchas His Pilgrimage*, sigs. N5V-N6R.

32 Drake, *Sacred Chronology,* sigs. C1R-V. See also William Birken, "Drake, Roger (1608–1669), physician and minister of religion."

33 Anon., "A Problem for Finding the Year of the Julian Period by a New and Very Easy Method," 324. De Billy's formula had first been recorded in the Parisian *Journal des sçavans* earlier in 1665.

34 Anon., "Why Is Wednesday November 17, 1858, the Base Time for VAX/VMS?" Digital Equipment Corporation-Customer Support Center. Colorado Springs. June 6, 2007. Archived from the original on June 6, 2007, accessed December 9, 2024. https://web.archive.org/web/20070606064704/http://vms.tuwien.ac.at/info/humour/vms-base-time-origin.txt.

35 J. D. North, "Chronology and the Age of the World," 307. On Ussher's calculations, in addition to the work of Barr (above), see also D. P. McCarthy, "The Biblical Chronology of James Ussher"; Michael Roberts, "Genesis and Geology UnEarthed"; Roberts, "The Genesis of John Ray and His Successors"; J. G. C. M. Fuller, "A Date to Remember: 4004 BC."

36 Alan Ford, "Ussher, James (1581–1656)."

37 Rudwick, *Earth's Deep History*, 14.

38 Rosenberg and Grafton, *Cartographies of Time*, 81. The title of the tables commemorated Kepler's patron, the Holy Roman Emperor, Rudolf II.

39 Jack Cunningham, *James Ussher and John Bramhall*, 89.

40 Ussher, *The Annals of the World*, sig. A4r.

41 See Ovid, *Metamorphoses,* trans. Raeburn, 9–12; Augustine, "Of the Catechizing of the Unlearned," 219. On ancient divisions of time, see E. J. Bickerman, *Chronology of the Ancient World.*

42 Milton, *Paradise Lost*, ed. Fowler, 26 (editor's note). All references to the poem are to this edition, hereafter cited parenthetically in the text by book and line number.

43 Ussher, *The Annals of the World*, sig. B1r.

44 Sprat, *The History of the Royal Society* ed. Cope and Jones, 161.

45 Anthony Hallam, *Great Geological Controversies,* 105.

46 Borst, *The Ordering of Time*, 103.

47 Stephen Toulmin and June Goodfield, *The Discovery of Time*, 141.

48 Volker R. Remmert, *Picturing the Scientific Revolution*, 3. Jessica Otis, *By the Numbers*, 2.

49 See Emily Grosholz, "Mathematics," 394–8.

50 Remmert, *Picturing the Scientific Revolution*, 3.

51 Rudwick, *Earth's Deep History*, 320. Modern geologists calculate time in units of a million years, known as a *mega annum* (Ma), and a billion years or *giga annum* (Ga). The BP timescale was introduced in the 1950s, with the advent of radiocarbon dating. Hence the use of January 1, 1950, as the "standard year" defining the "present."

52 Drake, *Sacred Chronology*, title page.

53 Douglas Bush, *English Literature in the Earlier Seventeenth Century*, 296.

54 By "semidiameter" Burton meant the distance between the center of the Earth and its circumference (i.e., its radius).

55 See Weija Wang, "Kant's Mathematical Sublime."

56 Anthony à Wood, *Athenæ Oxoniensis*, I. 627.

57 When, for example, Nathaniel Carpenter's geographical text entitled *Geography Delineated* was printed at Oxford in 1621 in an unusually large run of 1,250 copies, the 300-page book required some 200 reams of paper, or around 100,000 sheets. See I. G. Phillip, "A Seventeenth Century Agreement Between Author and Printer."

58 Locke, *An Essay Concerning Human Understanding*, ed. Nidditch, 207. Hereafter cited parenthetically in the text by page number.

59 See Sieghard Beller and Andrea Bender, "The Limits of Counting: Numerical Cognition Between Evolution and Culture." On different numeration systems and the linguistic concept of number, see David Crystal, *The Cambridge Encyclopedia of Language*, 15.

60 Archimedes, *Works*, ed. Heath, 232. See also *The History of Mathematics*, ed. Fauvel and Gray, 150–1.

61 The "Light Year" as a unit of astronomical distance, equivalent to the distance that light travels in a vacuum in 365¼ days, was first used in the 1860s. It is approximately equal to 5.88×10^{12} miles, or (as reported in 1890) 63,000 times the distance of the Earth from the sun. To associate the light year with the measurement of time rather than distance is sometimes taken as a sign of scientific illiteracy. See the *Oxford English Dictionary* (*OED*), "light year (n.)."

62 Bessemer, *An Autobiography*, 69. Bessemer was using the older (British) long form of a billion, which denoted one million million (i.e., 10^{12}) rather than the now widely adopted (American) short form, which defines a billion as a thousand million (i.e., 10^{9}). Today, Bessemer's calculation of a billion seconds would yield a result of 31 years, 251 days, 13 hours, 34 minutes, and 55 seconds. But the fact that either result seems plausible underlines the point that a billion is very hard indeed for the intellect to grasp.

63 Very large modern numbers possess the power to "tease us out of thought / As doth eternity" to recall John Keats's meditation on the passage of time which is his "Ode on A Grecian Urn" (1820), in *Poetical Works*, 210. The largest independently named number outside the Western decimal system is the Sanskrit *asaṃkhyeya* equivalent to 10^{140}. This term also means "incalculable." The googol (10^{100}) is the largest named number in the West, although the *OED* also cites the googolplex, defined as "ten raised to the power of a googol." See Eli Maor, *To Infinity and Beyond*, 16. The largest number discovered by contemporary mathematicians, and used in "a serious mathematical proof" is Graham's number. This number is so vast that a conventional digital representation would fill the observable universe, so that special notation has to be deployed to express the figure as follows: $3\uparrow\uparrow\uparrow 3 = 3\uparrow\uparrow(3\uparrow\uparrow 3) = 3\uparrow\uparrow 3^{27}$. See Martin Gardner, "Mathematical

Games," 28. I am grateful to Professor Bart for making me aware of (but not necessarily to understand) this number.

64 Willsford, *Willsford's Arithmetic, Natural and Artificial*, sig. C2r. For other examples of such popular arithmetic textbooks, see (e.g.,) Robert Recorde's much-reprinted *The Ground of Artes*; Humfrey Baker, *The Wellspring of Sciences*; and John Johnson, *Johnson's Arithmetic*.

65 Willsford, *Arithmetic*, title page.

66 See *OED*, "billion" (n. & adj.).

67 Morland, *The Urim of Conscience*, sig. C1v.

68 See *OED*, "myriad" (n. & adj.). In ancient Greek numeration, the word myriad (μυριάς) expressed numbers higher than 9,999. See Archimedes, *Works*, ed. Heath, lxix. The early modern English myriad is equivalent to the modern zillion or gazillion—a fictitious number denoting an indefinite quantity.

69 Morland, *A New Method of Cryptography*, sigs. A4[R-V]. Methusaleh was the longest living of the ancient patriarchs, dying at 969 years of age, according to Genesis 5:27.

70 The highest number represented in Hebrew, equivalent to the myriad, is *r'vavah* (רבבה), signifying 10,000, but also a large but indefinite number.

71 References to Shakespeare's plays are to *The Complete Works*, ed. Stanley Wells and Gary Taylor.

72 Blank, *Shakespeare and the Mismeasure of Renaissance Man*, 120–1.

73 Milton was recalling the angels, "beasts," and elders clustered around the heavenly throne in Revelation 5:11: "and the number of them was ten thousand times ten thousand and thousands and thousands" a figure which is indeterminate, but somewhere in excess of 100 million.

74 For comparison, Milton uses the word "millions" (always in its plural form) on eight occasions in *Paradise Lost*, while "thousand" or "thousands" occurs twenty times. See John Bradshaw, *A Concordance to the Poetical Works of John Milton*, 226; 364.

75 See Bradshaw, *Concordance*, 193. On Milton's use of numbers in *Paradise Lost* in the context of (late) seventeenth-century political arithmetic, see David Glimp, "Paradisal Arithmetic."

76 Lithgow, *The Present Survey of London and England's State*, sig. B2r.

77 Stephen Porter and Simon Marsh, *The Battle for London*, 7–8.

78 See Karl Pearson, *The History of Statistics in the 17th & 18th Centuries*, 30–40.

79 Thomas De Laune, *The Present State of London*, sig. E2r.

80 Wallis, *De sectionibus conicis*, sig. K2v.

81 See also Maor, *To Infinity and Beyond*, 7.

82 Augustine, *The City of God against the Pagans*, 432.

83 Ivano dal Prete, *On the Edge of Eternity*, 6; 11.

84 For a discussion of the distinction between the eternal and the sempiternal in the writings of Boethius and Thomas Aquinas, see Eleonore Stump, *Aquinas*, 131–58.

85 Vaughan, *Works,* ed. Martin, 466. See also Clark Hutton, "Images of Time and Eternity in the Religious Poetry of Henry Vaughan."

86 See Sharon Seelig, *The Shadow of Eternity*, 103–76.

87 Traherne, *Selected Poems and Prose*, 3.

88 Marvell, *Poems*, ed. Smith, 81–4.

89 Answers in Genesis, "Statement of Faith," accessed July 4, 2025, https://answersingenesis.org/about/faith/.

90 dal Prete, *On the Edge of Eternity,* 8–9.

91 Anon., *Creation Museum Signs*, 20. Various US courts have consistently found that "Creation Science" is not a "science," since its procedures are antithetical to scientific method. See, e.g., Michael Ruse, "Creation Science Is Not Science."

92 Anon., *Journey through the Creation Museum*, 27.

93 Answers in Genesis, "Statement of Faith," Accessed January 2, 2025. https://answersingenesis.org/about/faith/.

94 Anon., *Journey through the Creation Museum*, 31.

95 Answers in Genesis, "Statement of Faith," Accessed July 4, 2025. https://answersingenesis.org/about/faith. In using the term "kinds" the Creation Museum explains that "there can be many related species within a kind, but each kind is uniquely created and unrelated to other kinds." See Anon., *Creation Museum Signs*, 31. This is the main platform from which Young Earth Creationism has launched its attack against evolutionary modes of understanding.

96 See Rudwick, *Earth's Deep History*, 314–15.

Part Three

Poesy and the Mediation of Religion and Science

7

Spouse, Heir, and Handmaid: Composing Natural Philosophy in Bacon, Cowley, and Dryden

Mingna Cheng

Seventeenth-century England witnessed significant changes in the scope and value of natural science. The establishment of the Royal Society in 1660 marked a pivotal moment, characterized by the institution's pragmatic focus and transformative aim to advance knowledge. Francis Bacon, often considered the Society's intellectual progenitor, envisioned reformed modes of knowledge as capable of both comprehending nature and transforming society. This belief is evident in his choice of title, *Novum Organum* (1620), as a reference to Aristotle's work *Organum*, signaling a new instrument for acquiring knowledge. Bacon's emphasis on the practical applications of knowledge, epitomized by his assertion that "human knowledge and human power meet in one" (*scientia et potentia humana in idem coincidunt*), exerted a profound influence on subsequent English thinkers such as Thomas Hobbes and John Locke.[1] Hobbes, for example, echoed Bacon's instrumentalist philosophy in his own claim that "knowledge is power." He further articulated the purpose of philosophy as the practical application of knowledge to enhance human life, stating "the end and scope of philosophy is [...] to produce the like effects of those we conceive in our mind, for the commodity of human life."[2] In this vein, knowledge once vilified as the forbidden fruit leading to the Fall of mankind now emerged as the key to reclaiming dominion over the natural world. Though perhaps unintentional, seventeenth-century science witnessed an affirmation of human cognitive capacity and a consequent demand for a radical overhaul of epistemological foundations, ultimately shaking religious enchantment.[3]

Abraham Cowley and John Dryden, living inescapably under the shadow of the Baconian legacy, maintained close ties to the Royal Society, yet their attitudes toward the emerging scientific worldview diverged significantly. This essay contributes to ongoing scholarly discussions regarding the intersections of early modern science, poetry, and religion by examining how the nascent scientific paradigm influenced the poetic practices of Cowley and Dryden. Central to their differing poetic approaches is their conception of science's role in comprehending and transforming the natural world. Before delving into their poetic works, it is essential to outline Bacon's multifaceted epistemological project, upon which the Royal Society was founded.

Baconian Science Under Theology: The Snake in the Bosom

Early modern science did not begin as an enemy of religion.[4] Rather, while still in its infancy, it was known as "natural philosophy," a term that encompassed the pursuit of knowledge of nature and was deeply integrated within the broader intellectual framework of Christianity.[5] In the twelfth and thirteenth centuries, despite the dominance of theology in the culture of scientific inquiry, intellectual spaces emerged within newly established universities where pagan philosophy was studied. This period witnessed a renewed engagement with Aristotelian natural philosophy, fostering an environment conducive to theoretical and scientific inquiry. These rediscoveries did not necessarily lead to a clean break from Christian truth claims. Rather, scholars integrated pagan philosophy into a comprehensive Christian framework, positioning natural philosophy as a valuable, if subordinate, handmaiden to theology. This foundational relationship continued through the late medieval and early modern periods, even if it was continuously renegotiated. Within the Protestant Reformation's emphasis on secular labor and the progress of experimental inquiry, natural philosophy's status began to change.[6]

While medieval and early modern natural philosophy generally established its legitimacy within a theological framework, the grounds for this legitimacy began to shift. Often revered as a sacred text, nature was designated after Augustine as the "Book of Nature," providing the foundational path to divine encounter.[7] Nevertheless, a marked shift occurred in the goals and implications of seventeenth-century natural philosophy, particularly as articulated by Bacon. Bacon executed a radical inversion of Aristotle's epistemological hierarchy, elevating natural philosophy from Aristotle's "second philosophy" to the apex of knowledge (*philosophia prima*), "a parent or common ancestor to all knowledge" and "the great mother of the sciences" (3:346, 353, 4:78).[8] This reorientation positioned natural philosophy as a foundational discipline from which all other knowledge would spring. In contrast to the natural philosophy of antiquity and the medieval period, where the contemplation of nature was chiefly linked to human rationality, virtue, and the cultivation of the self, Bacon's vision increasingly viewed nature as primarily designed for human use.[9] Consequently, the study of nature became more instrumentalized, reimagined as the manipulation of a machine for human ends. Although Bacon defined the "history of Arts," which he called a discipline "of nature altered or wrought," as a branch of natural history, he criticized it as "deficient" because it had been handled "unprofitably" (3:330). Bacon in particular focused on practical applications rather than metaphysical speculations, laying the groundwork for the emergence of modern science.[10] Building upon the artisanal tradition of "maker's knowledge," Bacon's interventionist vision of natural philosophy redefined the human-nature relationship by integrating craft methods into a systematic framework for scientific inquiry and elevating practical knowledge. Such a view was enshrined in the proposed statutes of the Royal Society written by Robert Hooke: "the business and design of the Royal Society is to improve the knowledge of natural things, and all useful arts, manufactures, mechanic practices, engines, and inventions by experiments (not meddling with divinity, metaphysics, morals, politics, grammar, rhetoric, or logic)."[11]

As natural philosophy ascended as both a theological interpreter and a potent force for reshaping the physical world, it entered into a complex interplay with literature over the right to approach the Book of Nature. Poetry, as a linguistic art, has long been considered a means of representing or engaging with the natural world, though the precise method and extent of this engagement have been subject to continuous debate since Plato. Viewed as representation of reality, poetry prompted Plato to issue a warning about its potential to distance people from true knowledge, whereas the same view led Aristotle to assign poets with the task of learning universal truth through careful imitation.[12] Borrowing Aristotle's terminology as well as his optimism, Sir Philip Sidney compared poetry to a "speaking picture," treating it as a more reliable source of knowledge than history.[13] Most fundamentally, the Aristotelian concept of poetry as *mimesis* posed a critical challenge for seventeenth-century English poets: to what extent should poetic representation align with the newly emerging understanding of the world? Confronted by the growing influence and controversial claims of natural philosophy, poets were compelled to consider whether their art should respond to epistemic claims offered by the "new philosophy," a pursuit that some saw as a way for poetry to be "redeemed from error, from ignorance."[14] The inclusion of literary figures such as John Denham and Edmund Waller among the early fellows of the Royal Society, an institution that also sought to regulate and improve the English language, might seem anomalous without a discernment of the poet's duty to imitate nature.[15] For seventeenth-century intellectuals, however, both natural philosophy and poetry served as methodological approaches to theological exploration. The poetic controversies of the period emerged primarily from the challenge of reconciling poetry with the increasingly influential lens of natural philosophy in interpreting the Book of Nature.

Bacon's Many Faces

Although he is often cited as a figure who initiated the "separation between science and religion," in the words of R. F. Jones, Bacon did not view science and religion as fundamentally opposed.[16] Rather, he proposed a categorical distinction between "Divinity," grounded in divine revelation, and "Philosophy," derived from empirical inquiry that "spring[s] from beneath" (3:346). Concerned about the potential confusion between religious and scientific investigation, Bacon further subdivided philosophy into "Divine philosophy," "Natural philosophy," and "Human philosophy" (3:346, 349). This tripartite division, coupled with his characterization of natural philosophy as the "most faithful handmaid" of religion, underscores his view of these disciplines as complementary, rather than oppositional (4:89). Bacon's approach reflects a common attitude among early seventeenth-century English thinkers, who, following a long tradition of Christian thought, maintained that faith and reason, though separate, were ultimately in harmony, thus making the disciplines of religion and science "complementary to one another."[17] Through empirical investigation, Bacon sought to improve the human condition without compromising religious tenets. By exploring the Book of Nature, he believed humans could gain a deeper appreciation for the divine while simultaneously advancing human flourishing. Bacon's work thus inaugurated a new era of scientific inquiry characterized by a methodological distinction from religion, rather than a hostile separation.

Bacon regarded both his scientific and theological undertakings as having the same goal: to comprehend divine will. Employing a gendered metaphor, Bacon suggested that knowledge should not be merely pursued for pleasure or personal gain, but rather should serve as a generative and nurturing force:

> So the end ought to be, from both philosophies to separate and reject vain speculations, and whatsoever is empty and void, and to preserve and augment whatsoever is solid and fruitful; that knowledge may not be as a courtesan, for pleasure and vanity only, or as a bond-woman, to acquire and gain to her master's use; but as a spouse, for generation, fruit, and comfort. (3:294-95)[18]

Inductive philosophy, he argued, should forego speculative and fruitless pursuits in favor of activities that might produce more definite knowledge. In contrast to the prevailing view that knowledge is an end in itself, serving mere contemplation or personal gratification, Bacon asserted its instrumental value. He also outlined two "philosophies," one heavenly and the other earthbound. The former, rooted in contemplation and metaphysics, aligns with traditional natural theology, while Bacon grounds the latter on empirical inquiry and experimentation. It is this earthly philosophy, Bacon argued, that should inform the pursuit of knowledge as a generative force. Bacon's metaphor of knowledge as a "spouse" takes on additional significance in this context. For knowledge to fulfill its procreative potential, it must be coupled with the active, experimental method he advocated. By diligently investigating nature and applying the tools of empirical inquiry, humans could not only uncover the divine will as manifested in the natural world, but also leverage this knowledge to produce tangible benefits. Mere acquisition is insufficient; knowledge must be transformed into action, leading to human dominion over nature.

While Bacon presents his primary project of natural philosophy as generative in accumulating epistemological capital, he simultaneously relegates it to the subordinate position of the inferior second sex. This paradoxical construction, while often noted in discussions of Bacon's feminization of nature, is less frequently examined in relation to his portrayal of knowledge itself.[19] Bacon's conception of humanity's relationship to nature is similarly complex. His famous dictum that humans are "the servant[s] and interpreter[s] of Nature" establishes a foundational humility, extended by his analogy of natural philosophy as "the most faithful handmaid" to religion (4:47, 89). Bacon's natural philosophy, therefore, does not occupy a position of dominance or opposition to either nature or religion from its outset. Instead, it occupies a supportive and gendered role, one focused on the generation of new knowledge to serve human needs. The value ascribed to knowledge, in this framework, is primarily instrumental, emphasizing its productive capacity rather than its epistemological autonomy.

While Bacon assigned a crucial generative role to natural philosophy, likening it to a wife in a productive partnership with mankind, he did not envision a wholesale devaluation, replacement, or erasure of earlier methods of knowledge acquisition, such as those associated with poetry. Yet the question remains: to what extent did Bacon view poetry as antithetical to his scientific enterprise?

Bacon and the Limits of Poetry

While Bacon's passion for poetry undoubtedly paled in comparison to his scientific pursuits, he was not an acerbic critic of poetic arts in the manner of Isaac Newton, for example.[20] Drawing on a conventional taxonomy of human faculties divided by memory, imagination, and reason, Bacon correspondingly devised a tripartite classification of human knowledge: history, "poesy," and philosophy (4:292). Although acknowledging that poetry lacked the practical utility of natural philosophy, sharing the era's conviction of a profound connection between God, creation, and the Word (Logos), Bacon linked language intrinsically to the revelation of divine will and the comprehension of the cosmos. In *The Advancement of Learning*, for example, he defines "poesy" as a form of knowledge primarily constrained by metrical form but otherwise unrestrained and firmly rooted in the imagination. Recognizing poetry's exemption from the laws of the material world, Bacon nonetheless expressed reservations about its capacity for radical transformation (3:343).[21] Despite acknowledging poetry's widespread influence and popular appeal, Bacon ultimately confessed to having a more compelling intellectual interest in natural philosophy, famously declaring, "it is not good to stay too long in the theatre" (3:346).

Bacon's notorious critique of the "Idol of the Market-place" targets the abuse of language without condemning poetry.[22] As Jacqueline Cowan notes, for example, Bacon identifies a conflict between "poetic and rhetorical language," not between poetry and natural philosophy.[23] This distinction allows Bacon to exclude "verse" as mere "art of speech" while simultaneously affirming "feigned history or fables" as legitimate poetic forms (4:292). Bacon also categorizes poetry into narrative (historical verses), representative (visual history enacting natural events), and allusive (allegorical) modes, with the latter holding the greatest potential to "express some special purpose or conceit" (3:344). Refusing to elevate science as the sole arbiter of perception, Bacon acknowledges poetry's constructive and complementary role in interpreting the natural world through a mystical lens. The prevalence of parables and tropes in divine learning reinforces Bacon's assertion that sciences often require the support of similitude: "that whatsoever science is not consonant to presuppositions, must pray in aid of similitudes" (3:407). Elsewhere, Bacon notes that "there is no proceeding in invention of knowledge but by similitude" (3:218). This emphasis on similitude aligns with arguments that the Renaissance rhetorical tradition intervenes in knowledge acquisition. For example, Katharine Park claims that Renaissance man established his understanding of the physical world through "analogy" with knowledge he had gained from that "which is already understood," and it is upon this basis that the faculty of imagination could function as a "powerful, even necessary" instrument for the scientists.[24]

Thus, Bacon did not make a radical break with the poetic tradition, nor did he dismiss poetry's relevance to the pursuit of natural philosophy. His attitude toward imaginative literature evolved and added nuance over the course of his career. When he first published *The Advancement of Learning* in 1605, Bacon expressed satisfaction with the state of poetry: "in this third part of learning, which is poesy, I can report no deficiency" (3:346). But two decades later, in *Of Dignity* (1623), his perspective shifted. While preserving the value of certain ancient poetic works imbued with "mystic meaning," he criticized contemporary interpretations of these parables as inadequate (4:317-18). Besides, Bacon esteemed rhetoric as a "subservient" and useful instrument in assisting the faculty of imagination, and, even in the cost of a neat division, he wrote "the duty and office of Rhetoric [...] is no other than to apply and recommend the dictates of reason to imagination," affirming the poetic capacity to illuminate knowledge (4:455). Bacon's dissatisfaction lay not with the poetic form itself, but with prevalent misinterpretations ungoverned by proper judgment. This recognition led him to explore the potential of ancient fables to illuminate natural, political, and moral philosophy. For Bacon, both poetic art and natural philosophy were integral to the reconstruction of human knowledge. *The Great Instauration*, his ambitious project for human betterment, encompassed both material and spiritual renewal.[25] While natural philosophy demanded a rigorous, experimental approach, Bacon acknowledged poetry's unique capacity to transcend empirical limitations, reckoning it goes beyond the "measure of nature" in order to introduce "things which in nature would never have come to pass" (4:292). This imaginative power, he argued, was not a cognitive flaw but rather a distinct mode of understanding.

Bacon was not the only early modern author to praise the epistemological and representational functions of poetry. The English apologist George Puttenham, for example, elevated poets as "the first philosophers," attributing to them keen observation of "all natural causes and effect" and vehemently defending poetry against any form of disparagement by asserting its nobility, utility, antiquity, and divine nature.[26] Sir Philip Sidney also shared these sentiments, proclaiming the poet "monarch of all science" and asserting the superiority of poetry over both history and philosophy.[27] Sidney's vision was even more radical, positing poetry as a transformative force capable of elevating a "brazen world" into a "golden one."[28] Renaissance poets were thus perceived as "makers" possessing the creative power to shape both the reader's life world and an imagined realm through the potency of language.[29]

Bacon's departure from the Renaissance poetics of Puttenham and Sidney lies primarily in his skeptical and reflective stance on poetry's agency. One of the most distinctive features in Bacon's scheme of natural philosophy is the distinction between speculative and operative, the first dealing with the "inquisition of causes," the second with "the productions of effects" (3:351). Elsewhere, Bacon also asserted the fundamental distinction between artificial and natural things, stating that "artificial things differ from natural things not in form or essence, *but only in the efficient*" (1:496, my emphasis).[30] In this vein, we can infer that poetry, as the imitation of nature, belongs to the category of *artificialia* and thus is not in a position to affect the natural world. Unlike his contemporaries who endowed poetry with transformative power, Bacon regarded imaginative literature as lacking the capacity to alter nature in a fundamental way, since such nature operates according to its own internal laws.[31] While recognizing poetry's ability to represent the natural world, Bacon nevertheless viewed it as insufficient for the operative feature of natural philosophy. As Joanna Picciotto argues, embodying Adam as an innocent intellectual model anew, Bacon re-theorizes man's relation to the world by prioritizing "Adam's naming the creatures" and conferring on him the task of "working on" and "transforming" the world, claiming man's sovereignty "to be exercised over the creation itself."[32] This emphasis on human agency over nature aligns with Bacon's vision of a knowledge-driven transformation of the world. Bacon admits poetry's speculative capacity to understand the world (by means of similitude, for instance), yet he also identifies the limitations of his predecessors' understanding of it, arguing that poetry is futile in the operative attribute. This further inspires his pursuit of natural philosophy as a means to surpass what he saw as the limits of poetry and achieve knowledge's practical ends.

Bacon's *Instauratio Magna* not only led to the establishment of the Royal Society, it also inspired seventeenth-century poets who were sympathetic to the Society's mission. In his ode "To the Royal Society" (1667), for example, Cowley casts Bacon as a prophetic figure who led England from intellectual "barren Wilderness" to the "promised Land."[33] Similarly, Dryden in "To Dr. Charleton" (1663) hails Bacon as a pivotal figure between eras: "the world to Bacon does not only owe / Its present knowledge, but its future too," crediting him with both present and future knowledge (1:43). Both poets underscore Bacon's revolutionary role in overthrowing Aristotelian scholasticism. Dryden's epithet "the Stagirite" for Aristotle, who ruled for the "longest tyranny," was also notably used by Cowley in "To Mr. Hobbes" (1656) to emphasize the prolonged dominance of Aristotelian thought: "Long did the mighty *Stagirite* retain / the *Universal Intellectual Reign*" (2:19).[34] Yet Cowley and Dryden also diverged in their interpretations of Bacon's legacy, shaping distinct visions of natural philosophy. Dryden relegated natural philosophy to the subservient role of handmaiden in relation to nature, while Cowley elevated it to a position of prominence as the heir of all human knowledge. The following sections articulate how these poets appropriated and transformed Bacon's epistemological and metaphorical resources, revealing their divergent perspectives on the interplay between science and literature in interpreting the natural world in late-seventeenth-century England.

Cowley the Marcato

Cowley's attitude toward poetry underwent a significant transformation over the course of his life, profoundly shaping his philosophical outlook. Initially, as he suggests in "To Sir William Davenant" (1651), he embraced the notion of poets as "God-likc" figurcs who have the potential to "Heighten Nature by their verse" but do not alter such nature (1:144). This perspective aligns with Sidney's ideal of poets as refiners of the world and anticipates Dryden's later assertion in his *Defence of an Essay of Dramatic Poesy* (1668) that poets bring nature to a "higher pitch" (17:74). However, a more nuanced view emerges. As Gilliam Wright observes, Cowley's later work hints at "a more competitive relationship between Nature and Art."[35] While his elegy for Van Dyke subtly explores the tension between mimetic representation and the real world, with lines such as "Nature herself amazed, does doubting stand./ Which is her own, and which the Painter's Hand," his later "Second Olympique Ode of Pindar" (1656) offers a more direct critique (1:138). Here, Cowley situates art in a dependent and impoverished relation to nature, encapsulated in the lines "*Art* lives on *Nature*'s Alms, is weak and poor; / *Nature* her self has unexhausted store" (2:7). This evolving perspective culminated in a growing skepticism about poetry's capacity to represent the natural world accurately in his *Pindaric Odes*, which contrasts with his earlier and more optimistic view in the *Miscellanies*.[36]

Cowley's disillusionment with the limitations of poetic language coincided with his fascination with experimental inquiry and natural philosophy. In the 1656 preface to his *Poems*, for example, he explicitly repudiated his earlier poetic ambitions, deeming them the product of youthful infatuation with the "bewitching *Art*" of Poetry, confessing that he might find resolution in "those more profitable matches which I might have made among the *richer Sciences*" (1:cxxvii). Cowley's subsequent work reveals a marked shift toward philosophical and scientific pursuits as he sought to rectify poetry's perceived shortcomings. In "To Mr. Hobbes," Cowley resents how the old intellectual empire perished where "nought but Words it grew," and he performatively doubts the capacity of poetry to weave a suitable "Wardrobe" with which to "clothe the mighty Limbs of thy gigantic sense." Conversely, he extols Hobbes as the "great Columbus of the Golden Lands of new Philosophies," praising his mechanistic approach to understanding "Nature, and Causes" (2:19). Achsah Guibbory speculates that Cowley underwent a spiritual "conversion" in the late 1640s or early 1650s, marked by an increasing admiration for scientific figures like Hobbes.[37] Robert Hinman goes further, asserting Hobbes's "potent catalytic effect" on Cowley's thought.[38] In keeping with the mechanical philosophy of Hobbes, Cowley eulogized William Harvey's anatomical work for revealing the "mighty Mysteries" contained within nature, particularly by referencing his discovery of blood circulation. After referring to the capacity for poetic "art" to reveal "the heart of Man," Cowley turns, in an insinuating manner, to laud Harvey for reading "truth's own Book" instead of relying on "Comments only upon it"(1:158). Elsewhere, observing that poetry had been "usurped" for so long and so universally, Cowley writes poignantly: "There is not so great a *Lie* to be found in any *Poet*, as the vulgar conceit of men, that *Lying* is *Essential* to good *Poetry*" (1:cxxx). The impotency of "words" to fully grasp "things" becomes a recurring theme in Cowley's poems composed during the 1660s.

The relationship or hierarchy of words and things was a central concern for Cowley and his contemporaries in the Royal Society. Both groups shared a skepticism toward rhetorical embellishment, perceiving it as an obstacle to truth. Thomas Sprat, Cowley's biographer and literary executioner, explicitly advocated for a clear, direct style in scientific writing. Sprat urged readers to "separate the knowledge of Nature, from the colors of Rhetoric, the devices of Fancy, or the delightful deceit of Fables" and rejected "all amplifications, digressions, and swellings of style" in his apologetic work, *The History of Royal Society* (1667).[39] John Wilkins, Sprat's mentor, conceived an even more radical plan to establish a strictly one-to-one correspondence between the signifier and the signified under the aid of a "real character" and the associated "philosophical language" to avoid linguistic redundancy such as synonym or tautology.[40] Cowley himself agreed with these ideals of clarity and concision. In *A Proposition for the Advancement of Experimental Philosophy* (1661), he showed his growing disenchantment with the limitations of mimetic art. Unlike his earlier, more ambivalent stance, Cowley asserted in that treatise that poetry could not adequately capture or utilize the complexities of nature, declaring it is "like Painting by Memory and Imagination which can never produce a Picture to the life" (2:286). He contrasted the potential of experimental philosophy to "effect" knowledge with the perceived sterility of the ancient "solitary, inactive Contemplation of Nature" (2:285). This contrast reflects a broader intellectual movement away from Renaissance ideals of poetic creation toward a more empirical and utilitarian approach to knowledge. Cowley's work, in conjunction with that of the Royal Society, exemplifies this transition.

Cowley's title pays obvious homage to Bacon, yet he also modifies Bacon's categorization of human learning. In his preface, Cowley proudly announces that "all knowledge must be either of God, or of his Creatures, that is, of Nature," a departure from Bacon's previous trichotomy. In place of Bacon's three-part division, Cowley offers a dual division of knowledge into "Divinity" and "Natural Philosophy," which downgrades the status of history and "poesy" (2:285). In Bacon's taxonomy, history and poetry occupy equal status with philosophy in that the three correspond respectively to three human faculties (memory, imagination, and reason), whereas natural philosophy only performs as a sub-discipline of philosophy.[41] Yet Cowley raises this sub-discipline to the height of the first tier, and he further categorizes "Natural Philosophy" in relation to the object of study's proximity (immediate or mediate) to God. The study of God's indirect creations in such fields as grammar, law, and politics thus stands apart from the "mixed" investigation of God's direct creations, which encompasses the Book of Nature, humans, and other creatures (2:285). Cowley does not categorize poetry explicitly, but it is possible to conclude on the basis of his references to memory and wit that the mimetic power of that poetry falls into the category of God's immediate creation. Cowley also reveals his dissatisfaction with the existing knowledge concerning God's direct creation, lamenting how it is "slenderly provided for" or even "totally neglected" in the inquisition of nature, of which only mathematics and physics are, in an auxiliary manner, satisfactory in comparison (2:285). In this respect, Cowley's aspirations for the accomplishments and prospects of natural philosophy were even more ambitious than Bacon's.

In his Royal Society ode, appended as a prefatory poem to Sprat's 1667 *History*, Cowley reaches a crescendo of disillusionment with ancient epistemology. Deeply indebted to Bacon's metaphorical legacy, Cowley employs a striking analogy to highlight the inadequacy of ancient learning. By likening philosophy to a malnourished child, Cowley suggests that it is immature, underdeveloped, and incapable of fully comprehending or managing the complexities of the natural world. He lays blame squarely on these outdated intellectual traditions, arguing that they have left humanity ill-equipped to harness the resources and potential offered by nature:

That his own business he might quite forget
They amused him with the sports of wanton Wit,
With the Desserts of Poetry they fed him,
Instead of solid meats to increase his force:
Instead of vigorous exercise, they led him
Into the pleasant Labyrinths of ever-fresh Discourse:
  Instead of carrying him to see
The Riches which do hoarded for him lie,
  In Nature's endless Treasury,
  They chose his Eye to entertain
  (His curious but not covetous Eye)
With painted Scenes, and Pageants of the Brain.
(1:167)

Cowley here recalls Bacon's dismissal of ancient learning as the "boyhood of knowledge," highlighting its immaturity as captured in the phrase: "it can talk, but it cannot generate" (4:14). Bacon's longing for a natural philosophy capable of generative action resonates with a broader human desire for maturity and mastery. However, unlike Bacon, Cowley imbues his critique with a poignant regret for the neglected "Desserts of Poetry" and the enchanting "Labyrinths" of rhetoric.[42] Christening Bacon as a Moses-like figure and liberator, Cowley extols the founder of the Royal Society for leading humanity "From Words, which are but Pictures of the Thought [...] / To things, the Mind's right Object" (1:168). This emphasis on the tangible and empirical demonstrates Cowley and Bacon's shared conviction that language is susceptible to error. Yet, Cowley's admiration for Bacon's methodological rigor inadvertently extends to a misapprehension of Bacon's stance on poetry. While Bacon critiqued the abuse of rhetorical discourse, it is a stretch to equate this with a general distrust of poetry, a complex art form that combines language and imagination that might contribute to man's apprehension of the universe.

Cowley evinced a strong affinity for the institutions of linguistic renovation, notably expressed in his prefatory ode to Sprat's *History*. The ode displays Cowley's posture of intellectual struggle to find the equilibrium between the allure of poetry and the clear and secure comprehension of nature facilitated by science.[43] The composition of Cowley's prefatory poem was likely prompted by his prior reading of Sprat's *History*, indicating a connection between the two texts.[44] To establish a new kind of literary language, Sprat celebrated the benefits that experimental science could bring to literature, suggesting that by discovering "the Works of Nature" under the guidance of experiments, such science could become "an inexhaustible

Treasure of Fancy."[45] However, Sprat also reverses the complementary relationship between poetry and science originally envisioned by Bacon, relegating imaginative literature to a secondary and posterior position by suggesting that poetic creativity is contingent upon a deep understanding of nature gleaned through scientific inquiry. Likewise, whereas Sprat dismissed the voluminous "mists and uncertainties" that "Tropes and Figures have brought" to human knowledge, Bacon entertained the possibility of "Parables and Tropes" in "divine learning" (3:407).[46] Cowley, aligned with Sprat's anti-obscurantism, believed that "Nature's great Works no distance can obscure" and seeks to reconcile these divergent perspectives by positioning poets as fellow explorers of the natural world, striving to uncover the correspondence between phenomena and cosmic harmony (1:168). This vision emerges in his call for "good poets [...] who have purposely treated of solid and learned that is Natural Matters" (2:290) and achieves further definition in his own *De Plantis Libri VI* (Six Books of Plants, 1662–8). Cowley's pursuit of a "divine science" of poetry depends on his conviction that poets and natural philosophers share a common task: interpreting the Book of Nature.[47] By bridging the gap between poetry and natural philosophy, Cowley aims to create a new form of versification grounded in empirical knowledge, summarized by Samuel Johnson as an ideal of "improved and scientific versification."[48] It is perhaps the ultimate irony, however, that Cowley denigrated the very art of poetry in a poem, a paradox that mirrors Sprat's rhetorically sophisticated condemnation of tropes and figures.

Cowley's acute attention to language's enchanting power, initially channeled into poetry, eventually led him to embrace Bacon's pragmatic vision of natural philosophy as a generative force. Adapting Bacon's metaphoric framework, Cowley recasts natural philosophy as the "great and only Heir" of all human knowledge (1:167), a departure from the feminine, subordinate role of "spouse" assigned by Bacon (3:295).[49] This gendered re-imagining is significant. Cowley's heir is a male-coded term in that he endows this new philosophy with robust "Male-Virtue," and he manages to integrate Bacon's previous dismissal of ancient learning as an immature stage (still in "boyhood") and Bacon's designation of natural philosophy (the new learning) as the "spouse" at the same time. By hybridizing the two, Cowley not only realizes the generative role of natural philosophy but also, more importantly, establishes it as the rightful heir and owner of all other knowledge. This exclusive authority, epitomized in his determiner "only," enables natural philosophy to challenge scholastic orthodoxy ("put[ting] an end to their Authority"), renovate old learning, and dispel the illusions fostered by poetry. Cowley's metaphor of natural philosophy as "heir" signals his broader epistemological revision of Bacon, elevating natural philosophy to a status equal to divinity. By constructing natural philosophy as a privileged heir in terms of both gender and ownership, Cowley underscores its ascendance while simultaneously marginalizing the obsolete poetic mode.

Dryden's Decrescendo

Around the same time that Cowley composed the Royal Society's *Ode*, Dryden contributed his career-establishing poem *Annus Mirabilis* (1667), which would later earn him the title of Poet Laureate. This lengthy historical poem, which eulogized scientific and technological advancements, heralds both the scientific revolution and colonial empire.[50] However, recent scholarship suggests a more nuanced interpretation.[51] Despite being an early member of the Royal Society, Dryden tempers his enthusiasm for Baconian science with forms of skepticism.

In contrast to the anti-obscurantism shared by Cowley and Sprat, Dryden found himself drawn to the allure of mysticism and the supernatural, a domain that he believed natural philosophy, grounded as it was in rationality, could never fully encompass. In *Religio Laici* (1682), a poem composed during one of the most intellectually transformative periods of his life, Dryden pronounced a kind of death sentence on human reason under certain circumstances: "So pale grows Reason at Religion's sight;/ So dies, and so dissolves in Supernatural Light" (2:109). Reason has been accorded a secondary role in interpreting the divine scripture, a portion of which was manifested in the Book of Nature. It is from *Religio Laici* that Dryden, a gentry youth raised in a Puritan household, embarked on a trajectory that would culminate in his conversion to Catholicism years later. If we trace the origins of this development, we find an early manifestation in Dryden's juxtaposition of artificial capacity and natural wonder in the *Annus Mirabilis*, written in 1666. Following lavish praise for advancements in naval technology, Dryden writes in stanza 164: "Then, we upon our Globes last verge shall go, / And view the Ocean leaning on the sky" (1:84). Regardless of whether the Earth is flat or round, from a scientific perspective, it is perplexing to imagine how an ocean could lean against the sky or how a globe could possess "verge[s]." The editors of the California edition of Dryden's *Works* suggest that this stanza alludes to the cosmology of older astronomical models, where the terrestrial and celestial spheres meet at the horizon (1:301). In effect, the word "verge" allows Dryden to render explicit how Britain's transoceanic ambition for conquest arrived at its utmost limit. Nonetheless, in the meantime, it also implies that the progress of science can lead to an exaggerated and even absurd conception of nature, thus undermining the reliability and certainty of human reason. This implicit critique stands in stark contrast to Cowley's confident assertion of reason and experimentation as tools for understanding divine matters:

Yet Reason must assist too, for in Seas
So vast and dangerous as these,
Our course by Stars above we cannot know
Without the Compass too below.
(1:145)

Although Dryden's sense of reason, shaped as it is by a religious aura and supernatural light, appears somewhat diluted, Cowley demonstrates a more unwavering faith in reason as the dominant guide to perceive the "thousands" of lights in divine truth.[52]

Both Cowley and Dryden rely on metaphors of navigational technology—especially figures that invoke the magnetic compass that had made extended overseas exploration possible beginning in the sixteenth century, but they do so differently. Apart from emphasizing the compass as a guide for ocean voyage and imperial conquest, for example, Dryden also presents the absence of a compass and the primacy of natural forces in human engagement with the world in stanza 159:

Rude as their Ships was Navigation, then;
No useful Compass or Meridian known:
Coasting, they kept the Land within their ken,
And knew no North but when the Pole-star shone.
(1:83)

While Dryden ostensibly attributes the rudimentary state of navigation to the "Irish Kern" and "untaught Indian" (stanza 157), closer examination reveals a more nuanced critique of human technological hubris. The poet's implicit distrust of man-made instruments, such as the compass, is evident in his juxtaposition of them with the "Pole-star." This rhetorical strategy of transferring the looming consequences of dangerous techniques to the enemy while leaving the achievement to one's own nation allows Dryden to both extol the prowess of British maritime power and simultaneously express reservations about the reliability of human ingenuity (stanzas 162–6). The opening stanzas of *Annus Mirabilis* underscore this ambivalence. The catastrophic maritime disaster of 1666 in the Anglo-Dutch Wars, in which "human kind" is rendered powerless as "suffer[ing] Shipwrack everywhere" and "Heaven" would hide himself "in contempt of human skill" (stanzas 35–6), serves as a stark counterpoint to the later celebration of British naval triumphs (1:65). Dryden suggests that, despite scientific discoveries and technological advancements, humanity remains fundamentally subject to the capricious operations of nature. The pole star, a celestial guide, emerges as a counterbalance to the fallibility of human-made instruments, symbolizing a reliance on the natural order amid the uncertainties of human endeavor.

Rather than making an unbridled advocacy of the saving and transformative power of technology, Dryden underscored the limitations of it by invoking divine intervention. Later in the poem (stanza 281), the abrupt cessation of the Great Fire testifies to the supremacy of divine will over human endeavor. Dryden's depiction of the fire's extinction with a sense of *deus ex machina* suggests that, even in the face of such a catastrophic event, ultimate salvation lies beyond the realm of human agency:

> An hollow crystal Pyramid he [i.e. God] takes,
> In firmamental waters dipped above;
> Of it a broad Extinguisher he makes,
> And hoods the flames that to their quarry strove.
> (1:101)

Dryden's audacious juxtaposition of a celestial "crystal Pyramid" with a mundane "candle-snuffer" constitutes a quintessential example of metaphysical conceit. This incongruous pairing not only astonishes but also invites critical scrutiny. While Walter Scott aptly observes the bizarre employment of the extinguisher and the flames of London, which are "first a tallow candle; and secondly hawks, which, while pouncing on their quarry, are hooded with an extinguisher," the image also invokes a complex interplay between cosmic order and human artifice that reflects broader tensions between metaphysical and mechanistic tendencies in the period.[53] By elevating a quotidian object to celestial dimensions, Dryden challenges the emerging scientific worldview, which sought to reduce the universe to mechanical laws. The "crystal Pyramid" becomes a symbol of divine order, countering the mechanistic explanations gaining currency in the forthcoming age of reason. This deployment of metaphysics demonstrates Dryden's strategic use of poetry to resist the encroachment of scientific rationalism. His choice to adorn nature with the unexpected rather than subjecting it to the rigor of scientific inquiry aligns with his assertion in *Defence of an Essay of Dramatic Poesy* that poets possess a unique license to transcend the constraints of logic—that "it is not necessary for poets to study strict reason, since they are so used to a greater latitude than is allowed by that severe inquisition […] the poet dresses truth, and adorns nature, but does not alter them."[54] Remaining committed to poetic fancy, Dryden embellishes the natural world in terms of his metaphysical vision, drawing a conspicuous contrast with the new cult of Baconian science in his depiction of the aftermath of the Great Fire in stanza 256:

> Those who have none sit round where once it was,
> And with full eyes each wonted room require:
> Haunting the yet warm ashes of the place,
> As murdered men walk where they did expire.
> (1:98)

The haunting ghost of "murdered men" here compared to homeless fire victims offers a stark challenge to the mechanistic worldview that was gaining prominence in Dryden's time. The notion of a murdered man walking among the living defies the rational explanations promised by natural philosophy. Rather than eliminate such supernatural elements from his poetry, Dryden seeks a synthesis of the metaphysical and the mechanistic. Unlike Cowley, who advocated for a radical overhaul of poetic tradition, Dryden embraced the coexistence of such seemingly disparate perspectives. By preserving elements of the metaphysical, Dryden suggests that the poetic imagination offers a unique mode of understanding the world, one that complements rather than contradicts the insights of science. Acknowledging both the power and potential of natural philosophy, he remains committed to the enduring value of poetic imagination and its potential for both revelation and mystery.

Another distinctive feature of *Annus Mirabilis* is its significant debt to Virgil's *Georgics*, which, according to Dryden, is "the Divinest part of all his writings" (1:54-55). Michael McKeon astutely observes the prevalence of animal-related similes throughout the poem, such as places where Dryden compares the maritime war between England and the Netherlands to a clash between "elephants" and "rhinoceroses" or a "falcon" and a "crow" (stanzas 59, 86–7). Dryden reimagines Rupert's rescue at Albemarle in terms of a mother eagle protecting her young (stanzas 107–8) and portrays the Dutch as "the struggling Fly" ensnared by "the false Spider" (stanzas 180–1).[55] By adopting a Georgic and naturalistic lens, Dryden's extravagant metaphors diverge from the cold rationality of experimental scientists who often choose to distill their ideas through "Limbecks" (stanza 166; 1:85). Dryden's poetic imagination and literary fancy allow him to perceive and represent nature in a manner more akin to a poet than a natural philosopher. Such Georgic resonance evokes admiration and employs supernatural machinery to address the affairs of earth's creatures. It also echoes, in a conciliatory manner, Dryden's translation of *Georgics* in his later life (1697) in which he imagines Virgil as a member of the Royal Society: "Happy the Man, who, studying Nature's Laws, / Through Known Effects can trace the secret Cause" (5:203).[56] Dryden's *Annus Mirabilis* thus presents a fascinating dialectic, seeking to reconcile the poetic, Virgilian vision of nature with the emerging, mechanistic worldview of the scientific revolution.

Notably, Cowley also engages with Virgil's *Georgics* in his *Advancement*, including the work in the reading list prescribed to enhance the pupil's knowledge of natural history on the grounds that it "might serve for the advancement of Natural Science" (2:290). Cowley compiled a poetry collection that included verse translations from Virgil to Horace on husbandry and pastoral life. Cowley's intention is evident in the introductory essay he wrote for the anthology: "The first wish of Virgil (as you will find anon by his Verses) was to be a good Philosopher; the second, a good Husbandman," he writes, praising agriculture for its "necessity" to all "Mankind" (2:319-20). Dryden, in short, views Virgil's *Georgics* as a source of marvel and admiration while Cowley references Virgil's cultivation of nature in order to understand and apply concrete scientific principles.[57]

In contrast to Cowley, in other words, Dryden embraced the new science of the Royal Society without reservations about critiquing the traditional poetic approach, yet he still recognized the value and legitimacy of preserving it. Helen Burke astutely observes that, in writing *Annus Mirabilis*, Dryden responded to the intellectual tensions of his age, seeking "a similar reconciliation of conflicting natural philosophies" and thus "creating two different paradigms of man's relation to the natural world."[58] However, Burke's identification of Dryden's affinity to Sprat does not necessarily imply that Dryden's poetic aesthetics shares the same goals or objectives as Sprat's (or Cowley's) project to renovate poetic language. His poetic vision retains a certain layer of ambivalence and mystical overtones, which fundamentally influences his view of novel natural philosophy. Unlike Cowley, moreover, Dryden was less concerned with a crystal-clear understanding of the natural world. While Cowley aimed to integrate poetic art, a discipline heavily laden with similes and tropes prone to errors, into a clear, rational, and productive scientific linguistic system, Dryden's efforts were focused on reconciling the past and present simultaneously. Dryden's Janus-faced attitude ultimately led him to regard natural philosophy, along with other types of "Art," merely as "Nature's handmaid" (1:83).

In conclusion, both poets extend Bacon's legacy while maintaining distinct perspectives on the roles of poetry and science. Triggered by a growing distrust of words, as well as his intellectual exchanges with Hobbes,

Cowley focuses on the transformative and utilitarian promises of Baconian natural philosophy, advocating for a radical transformation of poetry itself. This emphasis leads Cowley to extend Bacon's critique of deceptive rhetorical discourse to the broader realm of letters, calling for a world governed by mechanism and empiricism. Cowley elevates the role of natural philosophy from its secondary position in Bacon as a "spouse" to that of the more equal heir—one entitled to inherit nature's resources for both the spiritual benefit and material betterment of humankind. In contrast, Dryden acknowledges Bacon's muted endorsement of "poesy" as imaginative literature, inferring a complementary role for poetry in understanding, interpreting, and appreciating the phenomenological world. Even within the scientific community, he implicitly advocates for the preservation of metaphysical extravagance and georgic naturalism in contrast to a world of physical certainty and mechanical clarity, offering a muted response to more enthusiastic choruses surrounding him in praise of seventeenth-century science. For Dryden, such science cannot help humanity to comprehend divine will, and even its generative promise is not always necessary. In certain circumstances, the Book of Nature, a proxy for divine existence, is to be simply marveled at rather than exploited.

Notes

1 Bacon, *Novum Organum*, in *The Works of Francis Bacon*, ed. James Spedding et al. 7:157. All references to Bacon's works are taken from Spedding's fourteen-volume edition, hereafter cited parenthetically in the text by volume and page number.

2 Hobbes, *The Metaphysical System*, 10.

3 For Max Weber's original account of the disenchantment thesis, see Weber, "Science as a Vocation," in *The Vocation Lectures*, 29–30. Admittedly, the academic popularity of the term "disenchantment" seems to be inversely correlated to its conceptual clarity, as it is often used "as a poetic synonym for secularization" (Jason Josephson-Storm, "Max Weber and the Rationalization of Magic," 31). Even concerted scholarly efforts, such as *Narratives of Disenchantment and Secularization* (2021), ed. Yelle and Trein, fail to provide a shared understanding of the term. In this article, I employ Weber's concept of enchantment to describe the pre-modern worldview where life is shaped by mysterious, supernatural, intractable, and ultimately uncontrollable powers, in stark contrast to the modern world's focus on rational understanding and human management. For a more detailed discussion of Weber's (dis)enchantment thesis, as well as its subsequent scholarship and criticism, see Nandini Das and Nick Davis, "Introduction,"

Enchantment and Disenchantment in Shakespeare and Early Modern Drama, 2–7.

4 It is crucial to recognize that the seventeenth-century understanding of "science" diverges significantly from its current usage. While modern science is primarily concerned with the empirical investigation of natural phenomena, the term "science" in the seventeenth century was more broadly encompassing. Derived from the Latin "*scire*" (to know), "science" (*scientia*) was synonymous with "knowledge" and could be applied to any field of study.

5 To adhere to period conventions, this chapter primarily employs the term "natural philosophy." Some argue that a clear distinction between "natural philosophy" and "natural science" did not emerge in Western thought until Isaac Newton or even Immanuel Kant, with the former term gradually yielding to the more specialized "science" of the nineteenth century (Moritz Schlick, *Philosophy of Nature*, 1; Lawrence Principe, *Scientific Revolution*, 27). Furthermore, in the Baconian sense, "natural philosophy" can also be referred to as the "new philosophy" (in contrast to the deductive old scholastic philosophy) or the "New Science," reflecting an aspiration to establish a novel epistemological framework. Occasionally, I use these terms, though I recognize nuanced differences between them.

6 For a more complete account of this change, see Weber, *The Protestant Ethic and the Spirit of Capitalism*, chapter 3. Robert Merton extends Weber's thesis to the world of the Royal Society in *Science, Technology & Society in Seventeenth Century England*.

7 Augustine coined this renowned expression in *Contra Faustum*, XXXII, 20 (*quasi legens magnum quendam librum naturae rerum*—"reading a kind of large book of the nature of reality"). See Augustine, *Answer to Faustus, A Manichean*, 422.

8 Aristotle, *Metaphysics*, 1037a15.

9 Crucially, however, this does not mean that the natural world lost its religious and moral meaning. For many seventeenth-century natural philosophers, the study of nature was seen as a religious duty—a way to restore humanity's dominion over creation, as given in Genesis (1:28; 2:15). As Peter Harrison has argued in *The Fall of Man and the Foundations of Science*, the emphasis on empirical and experimental methods was often conceptualized as a way to recapture the "Edenic wisdom," serving a theological purpose by compensating for the cognitive damage wrought by the Fall (190). Sorana Corneanu further adds that for Bacon, the pursuit of knowledge serves a dual purpose: it aims not only for the "benefit and use of man," but also for "the government and fortification of the mind" (*Regimens of Mind*, 33–4). Thus, theological and moral values were of vital importance to the practical goals of natural philosophy.

10 For relevant discussion, see Harrison, *Territories of Science and Religion*, 26–34, 117–44; R. G. Collingwood, *The Idea of Nature*, 3–9, 93–112.
11 Qtd. in Stephen F. Mason, *A History of the Sciences*, 259. On Bacon's debt to craft knowledge, see Antonio Pérez-Ramos, *Francis Bacon's Idea of Science*, and Cesare Pastorino, "The Philosopher and the Craftsman."
12 For relevant statements, see Plato, Book X of *The Republic*, and Aristotle, Chapter IX of *Poetics*.
13 Sidney, *An Apology for Poetry*, 10, 20.
14 John Dryden, "To Dr. Charleton," in *Works*, ed. Swedenberg, Jr. et al., 1:44. All references to Dryden's works, unless otherwise noted, are taken from this twenty-volume edition, hereafter cited parenthetically in the text by volume and page number.
15 Denham, Waller, and Dryden were among the original fellows. In addition, a host of literary men took part in the early inquisition of the Society, including John Evelyn and Cowley. For the inclusion of literati in the Society's early members, see Michael Hunter, "Analysis of Membership," 81–8. For the Society's literary aspiration upon its initial stage, see Hunter, *Science and Society in Restoration England*, 119.
16 Richard Foster Jones, *Ancients and Moderns*, 60.
17 Peter Zagorin, *Francis Bacon*, 49.
18 Bacon's connubial metaphor parallels the tradition of referring to the Church as the "bride" or "spouse" of Christ (Eph. 5:22-33; Revelation 19:6-8). A similar usage can be found in the works of John Donne: "Show me, dear Christ, thy Spouse so bright and clear" (*Complete Poems*, 464). Thus, an interesting analogy emerges when Bacon portrays natural philosophy as mankind's generative wife. Through this comparison, Bacon elevates and even sanctifies the role of man. I am indebted to Jen Rust for drawing my attention to this biblical trope.
19 Those discussions were first raised in the 1980s by three feminist historians of science, Sandra Harding, Carolyn Merchant, and Evelyn Fox Keller. For their main views, see Harding, *The Science Question in Feminism*; Keller, *Reflections on Gender and Science*, esp. 33–42; and Merchant, *Death of Nature*. These scholars all argued that Bacon's philosophy was sexist, accusing it of advocating the rape and torture of Nature. However, this standpoint has also been subject to attack and re-scrutiny in recent decades. For opposing opinions, see Iddo Landau, "Feminist Criticisms of Metaphors in Bacon's Philosophy of Science"; Brian Vickers, "Francis Bacon, Feminist Historiography, and the Dominion of Nature."
20 Newton was reported to describe poetry as "a kind of ingenious nonsense," though, as Richard Westfall noted in his biography of Newton, this remark may be apocryphal. See Westfall, *Never at Rest*, 581.

21 Being aware that poetry was "not tied to the laws of matter," Bacon continued to write about how poetry is inclined to "make unlawful matches and divorces of things."

22 "For it is by discourse that men associate; and words are imposed according to the apprehension of the vulgar. And therefore the ill and unfit choice of words wonderfully obstructs the understanding. Nor do the definitions or explanations wherewith in some things learned men are wont to guard and defend themselves, by any means set the matter right. But words plainly force and overrule the understanding, and throw all into confusion, and lead men away into numberless empty controversies and idle fancies" (Bacon 4:55).

23 Cowan, "The Imagination's Arts," 141.

24 Park, "Bacon's 'Enchanted Glass,'" 249.

25 Guido Giglioni points out the theological zealousness of "instauration." Bacon, as a faithful Protestant, believed that God's providence has decreed that "there should attend withal a renovation, and new spring of all other knowledges" ("Francis Bacon," 42).

26 Puttenham, *The Art of English Poesy*, 98–9.

27 Sidney, *An Apology for Poetry*, 25.

28 Ibid., 9.

29 See S. K. Heninger, Jr., *Sidney and Spenser*.

30 The Latin reads: "*artificialia a naturalibus non Forma aut Essentia, sed Efficiente solummodo, differre*."

31 "Nature does the rest by and within itself" (*ministrante sibi ipsi Natura*); see 1:497. The context in "De Augmentis Scientiarum" is Bacon's attack on the Renaissance notions of art as an amendment or perfecting of nature. For a more detailed analysis of how Bacon attacked human art as a discipline powerless to amend nature, see Giglioni, "Francis Bacon," 55.

32 Picciotto, *Labors of Innocence*, 2–3.

33 Cowley, *Complete Works*, 1:168. All references to Cowley's works, unless otherwise noted, are taken from this two-volume edition, hereafter cited parenthetically in the text by volume and page number.

34 A similar sentiment is also expressed by Joseph Glanvill: "A Schoolman is the Ghost of the Stagirite, in a Body of condensed Air" in *The Vanity of Dogmatizing*, 152.

35 Wright, "Nature in Cowley," 148.

36 It is hard to date the exact time of composition for every piece in Cowley's *Poems* (1656). However, according to Arthur Nethercot, the poems in *Miscellanies, Occasional Verses*, and *Pindaric Odes* were arranged chronologically. See Nethercot, *Abraham Cowley*, 48.

37 Guibbory, "Imitation and Originality," 107.

38 Hinman, *Cowley's World of Order*, 159. For Cowley's debt to Hobbes, see Chapters IV and V of the book.

39 Sprat, *History*, 62, 113. All citations of Sprat's *History* refer to the original 1667 edition.
40 See Wilkins, *An Essay towards a Real Character and a Philosophical Language*, especially the Fourth Part.
41 Medicine is another exemplary subject in Cowley's dismissal of what Bacon would designate as human philosophy (3:370), which Cowley places in the sphere of "all sorts of Natural, Experimental Philosophy" (2:288).
42 For Bacon's comparison of knowledge to food: "that knowledge is *pabulum animi* (the food of the mind)," see 3:383.
43 Wright, "Nature in Cowley," 146.
44 In a letter dated May 13, 1667, Cowley thanked John Evelyn for the compilation of potential subjects for a poem to be featured in Sprat's *History*, noting that he had already viewed certain sections of the text ("some parts of which I have seen"). See Evelyn, *Diary and Correspondence* (3:195-96).
45 Sprat, *History*, 413.
46 Sprat, *History*, 112; Bacon, 3:407.
47 Hinman, "Truth Is Truest Poesy," 200.
48 Johnson, *Lives*, 50.
49 Elsewhere, Bacon likewise declares moral philosophy to be "a humble handmaid" to divinity (3:433, 441). Harrison reminds us Bacon's metaphor is a reworking of the medieval commonplace of philosophy as handmaiden to theology. See Harrison, "Francis Bacon, Natural Philosophy, and the Cultivation of Mind," 146.
50 See Burke, "*Annus Mirabilis* and the Ideology of the New Science," and Stillman, "The State (Out) of Language."
51 For relevant research, see Joshua Brorby, "Seeming as Believing: Epistemological Uncertainty and the World of *Annus Mirabilis*," and my essay on, "Master or Handmaid?: Dryden's view of the New Science in *Annus Mirabilis*."
52 Richard Hillyer offers a more radical reading, arguing that the intended analogy only holds if the complementarity between "Stars" and "Compass" for mariners is equivalent to that of "The Holy Book" and "Reason" for Christians. However, he points out that celestial navigation would be impossible for sailors unable to distinguish between "numberless […] Stars" blending into "all one Galaxy" and virtually obliterating the night sky; they would have to rely solely on their "Compass." Thus, Cowley seemed to imply that Christians might be better off not blinded by an excess of "Lights," closing their "Holy Book" and relying exclusively on "Reason." See Hillyer, "Scientifick Versification," in *Four Augustan Science Poets*, 15.
53 Scott, commentary on *Annus Mirabilis*, in Dryden, Works, ed. Scott, 85.
54 Dryden, *Defence of an Essay of Dramatic Poesy*, in *Essays*, 1:121.
55 McKeon, *Politics and Poetry*, 154–5.

56 When translating Virgil, Dryden also introduced some contemporary nautical terms such as "Tack[ing] to the larboord," "Veer[ing] Star-board," and "weigh[ing] Anchors." Compare Dryden's *Aeneis*, in *Works*, 5:435, 471.

57 Bacon's use of *Georgics* appears in the construction of moral philosophy: "And surely if the purpose be [...] to instruct and suborn action and active life, these Georgics of the mind, concerning the husbandry and tillage thereof, are no less worthy than the heroical descriptions of Virtue, Duty, and Felicity. Wherefore the main and primitive division of moral knowledge seems to be into the Exemplar or Platform of Good, and the Regiment or Culture of the Mind; the one describing the nature of good, the other prescribing rules how to subdue, apply, and accommodate the will of man thereunto" (3:419). Bacon's potential debt to Virgil's *Georgics* is briefly explored in Anthony Low, "New Science and the Georgic Revolution," 249–55.

58 Burke, "*Annus Mirabilis* and the Ideology of the New Science," 309.

8

A “Goldmine of Inspiration”: Literature’s Role in the Science and Religion of Richard Dawkins and Thomas Sprat

Jacqueline L. Cowan

When Richard Dawkins attempts to bridge the divide between the humanities and the sciences, he echoes the rhetoric of a thinker that he might reject in other contexts. By proclaiming that “the poetry is in the science” in his best-selling book *Unweaving the Rainbow* (1998; 2000), Dawkins offers the poetic imagination as a new solution to the supposed incommensurability between literature and hard science.[1] Although Dawkins is an avowed atheist, the roots of his idea stretch at least as far back as the works of the English bishop and early Royal Society Fellow, Thomas Sprat. For both Dawkins and Sprat, recourse to poetry helped to resolve the religious tensions that plagued their forms of science. In both Sprat’s day and our own, scientists often appropriate the power of poetry to legitimize the authority and extend the purview of their respective sciences. The resurgence of Sprat’s line of rhetoric illustrates the inextricable ties among the fields of literature, science, and religion that today we consider discrete. By identifying literature as the thread that allows scientists to bring science to bear on religion, I show how literary scholarship can navigate the dense entanglements among early modern science, literature, and religion without succumbing to oversimplified and historically discredited conflict models.

Science, Literature, Religion, and Their “Conflicts”

For scientists and science popularizers such as Dawkins, science encompasses an objective truth about the natural world that is discovered empirically through a rigorously tested and repeatable method. For these thinkers, literature is a wondrous form of leisure rather than a way to make tangible meaning in and of the world. As Oxford chemist Peter Atkins claims, "although poets may aspire to understanding, their talents are more akin to entertaining self-deception. They may be able to emphasize delights in the world, but they are deluded if they [...] [believe this to be] enough for comprehension."[2] If Atkins is correct, the study of literature is a reflection or diagnostic of human culture at best. And, as Evelyn Fox Keller contends, this type of assumption about literature can exclude humanities scholars from setting foot into the territories of scientists.[3] Under this model, science and literature are closed off from one another, for its practitioners ply their trade in two incommensurable domains and toward incommensurable ends.

Religious studies, too, often finds itself overshadowed by the sciences. Dawkins's brand of atheism pits science against religious study and practice, claiming that a pervasive "God Delusion" (to borrow the title of his international bestseller) is equal parts irrational and dangerous.[4] Atkins goes further in claiming that science "should be acknowledged king" of all disciplines because it boasts an "omnicompetence" that renders obsolete other knowledge pursuits, particularly since "poetry titillates" and "theology obfuscates."[5] According to Atkins, science is staunchly opposed to literature and religion. And, until recently, scholars of early modern literature largely subscribed to these supposed oppositions, creating divisions where none have historically existed.

Both literary studies and the history of science once presupposed a staunch opposition between literature and science—and between science and religion. For literary scholars, this position is most clearly articulated in Denise Albanese's influential argument that "the emergence of modern scientific ideology in the seventeenth century resulted in the positing of fiction, of literary representation, as its binary (and prospectively devalued) opposite."[6] For historians of science, Peter Harrison notes that the "conflict model" between science and religion originated in the nineteenth century and has been perpetuated by "the assumption that science and religion are forms of knowledge based on mutually exclusive foundations."[7] For both literary scholars and historians, the conflict model argues that literature and science—and science and religion—worked counter to one another, each of its respective practitioners rejecting the aims of the other as antagonistic to their own. But more recent scholarship demonstrates that these conflict models erroneously address the developments of seventeenth-century natural philosophy—a field of study I discuss below—within the disciplinary boundaries that were only then coming to crystallize, mistaking science, religion, and literature as oppositional. In the process, conflict models risk retrofitting the seventeenth century with modern categories that obscure the mutually constitutive growth of, and entanglements among, these seventeenth-century fields of knowledge.

Within the last few decades, scholars of early modern science have largely abandoned models that understand science, religion, and literature as impermeable and oppositional disciplines. In literary scholarship, Mary Baine Campbell argues that "science and imaginative literature have made a dynamic pair of objects for study ever since they were sharply and categorically separated as activities of the mind and kinds of representations."[8] But she stresses the difficulty in positing such early modern boundaries because "poetry and science had not yet diverged and polarized."[9] Recent early modern scholarship addresses these permeable boundaries, channeling literary scholarship in new directions. Howard Marchitello, for instance, contends that there are two dominant historiographical narratives about the relationship between early modern science and literature, which he breaks into two waves: the first addresses the impact of scientific discoveries on literature and the second addresses literature and science as "equally (though, of course, differently) engaged in the production of knowledge."[10] More recently, Tita Chico goes so far as to argue that, in the eighteenth century, "science is a literary trope."[11] Historians similarly aim to move beyond any easy antagonisms between early modern science and religion. Harrison rejects the conflict model as an "erroneous view" and explains that "in so far as there is any general trend, it is that for much of the time religion has facilitated scientific endeavor."[12] Margaret J. Osler identifies several metaphors that structure the way scholars address the relationship between science and religion, noting that the notions of "conflict," "segregation," and even "harmony" all fail to capture the complexities of this dynamic early modern relationship.[13] She instead focuses on the "appropriation and translation of individual concepts from one discursive domain to another" (103), addressing how "natural philosophers [...] appropriate ideas developed in religion or theology, translate them into the language of natural philosophy, and use them to solve problems in the new context" (101). This approach to studying science and religion, Osler argues, "avoids the problems of essentialism that plague the more traditional models" and attends to the "permeable and elastic boundaries" of these fields of knowledge (103). I use these notions of appropriation and translation to move beyond the supposed antagonisms that animate Dawkins's work.

But even if scholars now embrace overlaps between science and literature or science and religion,[14] scholarship tends not to embrace the overlaps that inform all three fields of knowledge.[15] Instead, attempts to debunk the antagonisms that pit early modern science against literature and religion risk dividing them into separate camps: early modern scholars largely address either science and literature or science and religion, rather than all three at once.[16] As a stalwart critic of religion whose atheism is grounded in his understanding of science, Dawkins's focus on poetry reveals that these disciplinary tensions cannot be so easily parsed. Studying Dawkins's triangulation of science, religion, and literature helps break down barriers among literary studies and histories of science and religion, weaving together each strain of early modern scholarship. If scholars are to acknowledge the nuanced relationships between early modern science, literature, and religion, we ought to put them in conversation all together. I aim to contextualize historically the vexed relationship among science, literature, and religion in Dawkins's *Unweaving the Rainbow* to do just that. In the process, I hope to show how the tensions among these early modern fields of knowledge did not stem from any conflicting aims but out of competing claims for the same source of authority. Far from generating antagonisms, science, literature, and religion often shared commensurate goals and often performed commensurate work, albeit through different means and with different objects of study. Put differently, early modern science and literature did not conflict as two oppositional poles in a binary. Instead, their practitioners vied to align their studies with theology, each competing to show how their respective field of knowledge best reveals what they considered to be God's magnificence and, in the process, most effectively call humankind to virtuous action. So even if Dawkins mobilizes literature to support his argument that science ought to replace religion, his work evinces the mutual entanglement among these fields of knowledge, both in our day and in the early modern era in which we assumed they split.

Science and the Trials of Legitimacy

Thomas Sprat and his fellow inaugural members of the Royal Society mobilized the relationships among seventeenth-century literature, science, and religion to shore up the legitimacy of their scientific reforms. Much like scientists today, natural philosophers sought to discover knowledge of nature, albeit it through different methods and toward different ends. For centuries, natural philosophy was widely influenced by Aristotle, but the methods by which natural philosophers studied nature changed during Sprat's day. Sprat and other Royal Society Fellows advocated for a new natural philosophy that they called "experimental philosophy." This new philosophy departed in some important ways from the Aristotelianism that dominated natural philosophy: Royal Society Fellows relied on new instruments, such as the microscope, and on structured experiments to discover knowledge of nature. While we may recognize our own modern trust in new technology and experiments in these seventeenth-century experimental philosophers, the term "scientist" did not come into parlance until the nineteenth century.[17] Instead, the term "science" derives from the Latin word for knowledge itself, *scientia*. As Katharine Park and Lorraine Daston explain, *scientia* "referred to any rigorous and certain body of knowledge that could be organized (in precept though not always in practice) in the form of syllogistic demonstrations from self-evident premises."[18] It is difficult to avoid anachronism when discussing early modern natural or experimental philosophy as "science," and not just because of our terminology. Because modern science is widely acknowledged as the most reliable source for knowledge of the natural world, it is easy to forget that "science" once struggled to validate its practices in various circles of orthodox authority. As Carla Mazzio argues, "'Science' as a term has now accrued meanings and forms of cultural capital that would have made Shakespeare's eyeballs start from their spheres."[19] It is important, then, to address seventeenth-century experimental philosophy, poetry, and theology on their own terms.[20]

In the seventeenth century, poetry commanded significant cultural authority—an authority so powerful it often sparked fears over poetry's uses. Concerned about the misuses of poetic language, for example, Stephen Gosson wrote *The School of Abuse* (first published in 1579), a Puritanical attack that, as its full title proclaimed, "contains *a Pleasant Invective Against Poets, Pipers, Players, Jesters,* […] [and raises] *Up the Flag of Defiance to their Mischievous Exercise.*"[21] In this pamphlet, Gosson acknowledges that poetic language, whether on the page or on the stage, has the power to move an audience to action. Yet, when mishandled, this power might provoke all manner of ill deeds: "I set down of the abuses of Poets, Pipers, and Players which bring us to pleasure, sloth, sleep, sin, and without repentance to death and the Devil."[22] Although Gosson himself was a playwright, *The School of Abuse* emblematizes the concerns over poetry that Philip Sidney sought to allay in *The Defense of Poesy* (1595). While Sidney notes that "man's wit" can "abuseth poetry," he argues that poets are best at inspiring virtue, which he understands as the "end of all earthly learning."[23] For Sidney, the poet stands above the historian, who is too tied to particular events to reveal universal truth, and the philosopher, who is "so hard of utterance" and "so misty to be conceived" (221) that only a "learned" handful may understand what is being taught (223). The poet, however, is able to "teach and delight" (217), rendering moral values palatable and easy to understand for diverse audiences: "to move men to take that goodness in hand […] [and] to make them know that goodness whereunto they are moved" (218). Counter to Gosson's claims, Sidney details the ways that poets serve goodly ends—they "lead and draw us to as high a perfection as our degenerate souls [...] can be capable of"—and do so with greater success than other vocations (219). Although the poet's place in society was not uncontested, poets were thus understood to hold significant cultural clout because they could call humankind to action, be it sin or virtue, in ways other professions could not. And this was clout that, as we shall see, Royal Society Fellows aimed to appropriate.

Sidney adds "natural philosopher" (216) to the list of professions over which the poet ought to be heralded as "the monarch" (226). For Sidney, the poet can imitate the natural world and, more importantly, make it serve didactic ends; the natural philosopher, however, must represent the natural world only as it is. The poet alone, Sidney explains, "goeth hand in hand with nature" but "disdain[s] to be tied to any such subjection, [and so is] lifted up with the vigour of his own invention [...] in making things either better than nature bringeth forth, or, quite anew, forms such as never were in nature". The poet's fictional world surpasses the natural world that the natural philosopher studies because it is able not only to call humankind to virtuous action but also to reveal the magnificence of God in sharper relief: "Nature never set forth the earth in so rich tapestry as divers poets have done [...]. Her world is brazen, the poets only deliver a golden" (216). As S. K. Heninger, Jr., understands it, Sidney claims that the poet "creates 'another nature,' a meta-reality, a world closer to God's design than the corporeal universe because it transcends the partiality and accidents of particular instance."[24] In their capacity to study only the "brazen" world, Sidney argued that natural philosophers remained beneath poets in status.

Theology, too, had a higher status than natural philosophy.[25] Religion influenced all aspects of early modern England. As David Cressy and Lori Anne Ferrell explain, "the pattern of the cosmos, the history and destiny of the world, the ordering of social, political, and domestic relationships were all explained in biblical and theological terms."[26] Rivka Feldhay argues that while medieval Christian culture "absorbed" Aristotelian natural philosophy and other "classical Greek bodies of knowledge," various "cultural mechanisms" kept this scientific knowledge in check. These mechanisms, she continues, included an epistemological hierarchy with "Christian theology lodged securely at its apex" and the institutional "creation of a class of Church intellectuals—mainly theologians—who were responsible for the selection, legitimization, use, and suppression of other bodies of knowledge in accordance with Christian concerns."[27] Harrison, however, contends that, at least for Aristotle, this hierarchy was more than just a way of classifying different types of knowledge. It "set out the proper progression in the order of study, beginning with the material and mutable (natural philosophy)" and culminating with the "elevated divine things that were immutable and completely divorced from matter."[28] Under this hierarchy, natural philosophy was an important preparative for theological study—a step that could not be skipped lightly. Rather than "breaking free of religion," Stephen Gaukroger argues that "Christianity took over natural philosophy in the seventeenth century, setting its agenda [...] and in the end establishing it as something in part constructed in the image of religion."[29] When Royal Society Fellows began to advocate for their new experimental philosophy, they had to cultivate unique relationships with the reigning authorities over knowledge: the poet, and, more importantly, the theologian.

And Royal Society Fellows faced more than an uphill battle. Samuel Butler deflates the pomp and hubris he attributes to Royal Society rhetoric, reducing the "VIRTUOUS, learn'd *Society*, of late / The Pride and Glory of a foreign State" to a bunch of buffoons who mistake a mouse trapped in a telescope tube for a lunar elephant.[30] Margaret Cavendish lambasts the Royal Society by transforming them into anthropomorphic "bear-men" who "take more delight in artificial delusions, than in natural truths" in her *Description of a New World, Called the Blazing World* (1666).[31] In *Gulliver's Travels* (1726), Jonathan Swift satirizes the Royal Society in more scatological terms, tasking the "*grand Academy of* Lagado" with turning feces back into human food.[32] Fellows also bore the ire of churchmen but often on practical rather than theological matters.[33] Much like the poets, these ecclesiastical critics claimed that experimental philosophy was fruitless or inane. Clergyman Robert South, for instance, scoffed that the Royal Society's use of the microscope rendered their work idle as they "can admire nothing except fleas, lice, and themselves."[34] Royal Society Fellows faced criticism from a variety of different and influential early modern thinkers. As Harrison points out, "the initial attempts of Royal Society apologists to vindicate experimental philosophy elicited considered counterarguments that went well beyond the casual mockery of its more popular detractors" (*Territories*, 128). In response, Royal Society Fellows such as Sprat worked tirelessly to advocate for new experimental methods, smoothing over any contentions with the authorities whose approval they sought.

The crisis of authority that Dawkins's science faces in our contemporary world pales in comparison to the one that Sprat worked to overcome. But both make recourse to the poet to grant science greater authority. Dawkins's *Unweaving the Rainbow* is a riposte against the claim that science denudes the world of meaning beyond its material substrates. Far from a means to disenchant the world or, as Keats puts it, "unweave a rainbow," Dawkins argues that science fuels the imagination, drives us to seek new knowledge of nature, and ultimately reveals a wondrous world worthy of awe.[35] Poetry is wholly severed from the natural world that occupies the studies of science, but can whet the appetite for wonder and, in the process, extend the search for knowledge forever forward. "The feeling of awed wonder that science can give us," he writes, "is one of the highest experiences of which the human psyche is capable. It is a deep aesthetic passion to rank with the finest that music and poetry can deliver" (*Rainbow*, x). Dawkins doubles down on this argument in his most recent book—a collection of influential lectures and writings entitled *Science in the Soul* (2017). Here he claims that "it's high time the Nobel Prize for Literature was awarded to a scientist."[36] Dawkins flags "real poets and true scholars of literature" as science's great cheerleaders since they are "the very people best qualified to appreciate [the wonders of science] and inspire others." Dawkins quotes Carl Sagan as a great advocate of the "true poetry of science," identifying poetry as the lynchpin between science and religion. Sagan argues that "a religion, old or new, that stressed the magnificence of the Universe as revealed by modern science might be able to draw forth reserves of reverence and awe hardly tapped by the conventional faiths."[37] Both of these influential scientists use the poetic imagination to bring science squarely to bear on religion: the enchantments of religion can take a back seat to the poetic wonders of science. Dawkins thus advocates for a closer relationship between poets—whom he considers the traditional source of wonder and enchantment—and scientists, whose understanding of the material world he thinks should merit aesthetic appreciation. Dawkins summarizes his vision for poets and scientists:

> It is my thesis that poets could better use the inspiration provided by science and that at the same time scientists must reach out to the constituency that I am identifying with, for want of a better word, poets. It is not, of course, that science should be declaimed in verse [...]. Simple, sober clarity will do nicely, letting the facts and the ideas speak for themselves. The poetry is in the science.
>
> (*Rainbow*, 17–18)

Dawkins's late-twentieth-century "thesis," however, is well over three centuries old.

Dawkins's Debt to Sprat and the Poetry of Science

In his *History of the Royal Society* (1667), Thomas Sprat made the same argument in an attempt to establish the Royal Society's notion of experiment-based natural philosophy as the dominant authority over knowledge of nature.[38]

Echoing Francis Bacon, Sprat's *History* argued that this new philosophy would bring about an age of discovery, driving knowledge into earthly and celestial domains never before reached, and all for the benefit of humankind. But such discoveries, Sprat claimed, would be obscure if delivered through ornate, elevated language; poets could no longer wear the crown Sidney offered them many decades earlier. Sprat proclaims that "*eloquence*," here rhetorical and poetic language, "ought to be banish'd out of all *civil Societies* as a thing fatal to Peace and good Manners" (notwithstanding his own use of poetic language, of course) (111).[39] Instead, Sprat advocates for "a close, naked, natural way of speaking" (113) that Dawkins echoes when he asks for "simple, sober clarity" from the language of science (*Rainbow*, 18). Only then, Sprat and Dawkins intimate, can science progress without impediment.

For Sprat and his experimental philosophers, the poet was a competing claimant on the knowledge of nature. But, as the many failed critics of poetry before Sprat evinced, poetic language cannot be successfully divorced from science nor from society. Royal Society Fellows thus proposed a new tactic to diminish the poet's authority: to surpass poetry on its own terms. Far from meddling with a "brazen" natural world, as Sidney earlier argued, experimental philosophers claimed to discover in nature a world more beautiful and more magnificent than whatever the poet could pen. Founding Royal Society Fellows claimed the "golden" world as their own domain. For instance, Robert Hooke, the early Royal Society's most famous microscopist, marvels at the wonders of the "new visible world" revealed through the experimental philosopher's magnifying lens:

> *By this means* [of the telescope] *the Heavens are opened, and a vast number of new Stars, and new Motions, and new Productions appear in them, to which all the ancient Astronomers were utterly Strangers. By this* [microscope] *the Earth itself, which lies so near us, under our feet, shows quite a new thing to us, and in every* little particle *of its matter, we now behold almost as great a variety of Creatures as we were able before to reckon up in the whole* Universe *itself.*[40]

The Royal Society's new instruments uncovered golden worlds; their new natural philosophy was not anathema to wonder but the primary means by which to produce it. To borrow Dawkins's turn of phrase, "the poetry is in the experimental philosophy."

By revealing a wondrous "golden" world in nature, experimental philosophers claimed to trump the poet's imagination. For, as Sprat argued, poetry is marred by a "peculiar *imperfection*, that they were only *Fictions* at first: whereas *Truth* is never so well expressed or amplified as by those

Ornaments which are *True* and *Real* in themselves" (414). Securely tethered to the material world, experimental philosophers flipped the logic by which Sidney enthroned the poet as the "monarch" of the sciences: all of the wonders that Sidney thought revealed by grace of the poet's imagination were now by nature made privy to the experimental philosopher alone. Royal Society Fellows proclaimed that when it came to addressing the magnificence of nature, poetry was obsolete in the face of the new experimental philosophy. Sprat perhaps best articulates the poet's obsolescence when he claims that "the *Wit* of the *Fables* [...] of the *Ancient World* is well-nigh consumed: They have already served the *Poets* long enough, and it is now high time to dismiss them" (414). By arguing that poetry was obsolete in the service of beautifying nature, Royal Society Fellows co-opted the poet's authority as their own. Early modern poetry and experimental philosophy, then, were not understood as antagonistic because they did not work toward incommensurate or oppositional ends. On the contrary, they both performed similar work: they uncovered the divine imprints in nature, inspired wonder, and, as I discuss below, called humankind to virtuous action. But because experimental philosophers plied their trade in the "real" world, they could differentiate their work from the poet's and claim its superiority.

Herein lies the connection between Sprat and Dawkins and the central irony of Dawkins's line of rhetoric. Both Sprat and Dawkins couch their defense of science in the rhetoric of poetic wonder.[41] Dawkins's recourse to the language of "wonder" allows him to bridge science and literature: science does not render nature a bald, boring object of study but whets a sense of poetic awe in any who dare explore nature's magnificent treasures. But his attempt to draw poets and scientists into greater alignment repeats the very argument that divided them over three centuries ago. When Dawkins asks science to call upon the poet's imagination, he cannot help but subordinate poetry under the scientist's authority in the same way that Sprat does. Both Dawkins and Sprat are leery about the deleterious effect poetry may have on scientific knowledge. Dawkins warns against "bad poetic science," by which he means meaningless or aberrant metaphors of pseudo-science that siphon off poetic wonder for fruitless mystical forays (*Rainbow*, 180). Sprat too laments the "mists and uncertainties [...] [that] specious *Tropes* and *Figures* have brought on our Knowledge" (112). For both Dawkins and Sprat, an overactive fancy paired with unruly poetic language leads to superstitions that impede science and halt the search for truth. Dawkins debunks "bad poetry" on the same grounds that Sprat decries the poet for crafting false images that he thinks prevent humans from celebrating the divine hand that created nature. Science and literature are not binaries and never have been, but their practitioners did and do still compete for authority on similar terrains. For this reason, both Dawkins and Sprat suggest that science best engages with the poet when the poet writes about science.

For Dawkins and Sprat, science lifts nature's hermetic lid to reveal a treasure trove of wonder that spurs the imagination. Sprat proclaims that the "*Works of Nature*" house an "inexhaustible Treasure of *Fancy*," but this is a trove of wonders that only experimental philosophers can open (413). By revealing nature's "Treasure of *Fancy*" Sprat argues that experimental philosophers provide fodder for poetic endeavors: they discover "very many *new things* to be named, and adorned, and described in [the] discourse" (324) of both the poet and rhetorician and that these "discoveries will be very serviceable to the *Wits*, and *Writers* of this, and all future *Ages*" (413). Sprat assures the poet that these discoveries will not impede poetry because experimental philosophers simply add more materials to whet the poet's imagination. If perhaps by this means, any change shall be made herein," Sprat contends, "it can be only for the better" (324). Dawkins echoes Sprat's argument about nature's "inexhaustible Treasure of *Fancy*," claiming that poets "have overlooked the goldmine of inspiration offered by science" (*Rainbow*, 15). Dawkins similarly argues that any poet who casts aspersions on science would be a "waste of poetic talent" (*Rainbow*, 17). In *Unweaving the Rainbow*, most of Dawkins's chapters delineate the wonders discovered by science, including the awesome beauty of the rainbow's wavelengths, of human DNA, and of the human body and brain. Poets, Dawkins urges, are remiss to ignore such fruitful sources of inspiration.

For Sprat, poets could continue their work under the auspices of the experimental philosophers who would inspire their labors. He concedes that experimental philosophers could:

> commend [those that use poetic language], as we are wont to do *Chaucer*; we would confess that they are admirable in comparison of the ignorance of their own Age. And, as Sir *Philip Sidney* of him, we would say of them; that it is to be wondered how they could see so clearly then, and we can see no clearer now.
>
> (21)

But, Sprat stipulates, those who ply their trade in eloquent language "ought to be confined within their own Bounds, and not be suffered to hinder the enlargement of the territories of other *Sciences* […] let them not over-spread all sorts of knowledge" (21). Royal Society Fellows rendered poetry obsolete as a science at the same time they claimed to provide the poet with new scientific muses. Because Dawkins delineates science's many wonders in such a successful and exhaustive fashion to disprove Keats's argument that "cold [natural] philosophy" can "unweave a rainbow," Dawkins too subordinates the poet to the scientist's authority.[42] Like Sprat, Dawkins takes great pains to identify why poets ought to take science as their inspiration. But he is less clear on how, as he argues, "scientists must reach out to the constituency [...] [of] poets." Dawkins staunchly contends that "the poetry is in the science," but one imagines that he would be far more reticent to suggest that the "science is in the poetry." Dawkins repeats the same line of rhetoric that early Royal Society Fellows did well over three hundred years ago, and the result is much the same. When Dawkins argues that the "poetry is in the science," he thus seizes the domain that he originally ascribes to poetry for his science. Experimental philosophy for Sprat and science for Dawkins co-opt the poetic imagination, summarily denying poets entry into the now enlarged domain of scientific authority. Both Sprat and Dawkins bring the new poetic quality that they ascribe to their respective sciences to bear on religion but to opposite ends.

Weaving Together Science, Literature, and Religion

Far from promoting conflict, Sprat worked diligently to align the new science with religion; he considered his "defence of the *Royal Society*, and this new *Experimental Learning*, in respect of the *Christian Faith*" to be the "weightiest and most solemn part of [his *History*'s] whole *undertaking*". For Sprat and other Royal Society Fellows, religion was a far more "slippery place" than was the terrain of poetry because "it is almost impossible without offence, to speak of things of this Nature" (345). Both Sprat and Dawkins work to curb what Sprat understands as the threat of religious "*Enthusiasm*" (428) and what Dawkins considers religious "fundamentalism" and "fanaticism."[43] Although Sprat was a devout member of the Anglican Church and Dawkins is a stalwart atheist, each worries about the deleterious effect of inflammatory rhetoric and religious passion. Sprat writes in the wake of the bloody English Civil Wars and, fearing future turmoil, argues that experimental philosophy would extinguish the flames of civil unrest. This turmoil was largely blamed on "enthusiasm," which Michael Heyd argues was the "standard term used to designate individuals or groups who claim to have direct divine inspiration."[44] Sprat argues that because experimental philosophers used plain language, the disputes that could fan larger tensions were avoided. Further, he contends, the "pleasure" of experimental "*Discoveries* [...] will wear off the roughness" of temperament that leads to "*Rebellion*" (428). Here Sprat assures his detractors that the new experimental philosophy will help quell "the fury of

Enthusiasm" (428). Dawkins, too, aims to legitimize his science by framing it as an antidote to the "genie of religious fanaticism" and fundamentalism (*Delusion*, 63). But where Dawkins turns to science to promote atheism, Sprat turns to experimental philosophy to align the study of nature with theology.

It is important to note again that experimental philosophy in Sprat's era was not in conflict with theology; many scholars have demonstrated how experimental philosophers worked to legitimate their new practices in religious spheres. Angelica Duran, for instance, points to the ways in which "New Scientific texts" adopted the "format of the catechism" to explain "the process by which modern science gained authority so powerfully and so quickly."[45] James J. Bono has also turned to Robert Boyle in order to demonstrate the inextricable ties between experimental philosophy and theology, arguing that "*how* to read the Book of Holy Scriptures also lays the foundation for *how* postlapsarian humans must learn to read the Book of Nature."[46] As Courtney Weiss Smith makes clear, moreover, "the book metaphor was not just a hackneyed convention allowing scientists to gesture politely to God's design before they got on to the work that really interested them," even if this "is how scholars today often understand the new science's treatment of it."[47] Harrison surveys the relationship between experimental philosophy and theology more broadly, arguing that "part of [the Royal Society's] strategy was to appeal to religion, the legitimacy of which was at that time beyond question" (*Territories*, 126). Indeed, as Amos Funkenstein so influentially argues, "science, philosophy, and theology [were] seen as almost one and the same occupation" for parts of the sixteenth and seventeenth centuries.[48] Sprat's concerns about experimental philosophy's relationship to Christianity thus do not indicate any antagonism between early modern science and religion. Instead, they bespeak the Royal Society's attempt to secure for experimental philosophy the same close relationship between natural philosophy and religion. Sprat even professes his gratitude to those religious clerics who defend experimental philosophy:

> Of our *Churchmen* the Greatest and the most *Reverend,* by their care, and passion, and endeavors, in advancing this *Institution,* have taken off the unjust scandal from *Natural* knowledge, that it is an Enemy to *Divinity.* By the perpetual *Patronage* and *assistance* they have afforded the *Royal Society,* they have confuted the false opinions of those men who believe that *Philosophers* must needs be *irreligious:* they have shown that in our *veneration* of *God's almighty power,* we ought to imitate the manner of our respect to *Earthly Kings.*
>
> (132)

The Royal Society buttressed their commensurability with theology by claiming to make humans more virtuous by studying nature in ways that promoted the ends of religion. According to Sprat, both tactics were secured by the poetic power that experimental philosophers wielded. Experimental philosophers thus used their new relationship to poetry to outmaneuver, at least in part, attacks against experimental philosophy.

Against the criticisms lodged by poets and churchmen, Sprat rejects the idea that experimental philosophy is an idle pursuit and declares that the Royal Society works for "the Benefit of human life" (2). As scholars have shown, the Royal Society advertised the practical applications of their studies and argued that their discoveries would regain, to the highest degree possible, the Edenic relationship between Adam and nature before the Fall.[49] For Royal Society Fellows, such a golden age could not rest on armchair philosophy. As Sprat contends, the Royal Society eschewed the idle debates of the "*School-men*" who are "tied by their Cloistral life to such a strictness of hours, and [who] had seldom any larger prospects of *Nature* than the Gardens of their *Monasteries*" (19). Sprat thus welcomes the "admission of Men of all *professions*" into the ranks of other gentlemen of their society (65). Unlike the scholastics and the poets, Sprat contends, experimental philosophers worked in the "real" world. For Gaukroger, this shift moves natural philosophy away from "detachment and contemplation (*otium*)" and squarely toward the "affairs of state (*negotium*)," allowing figures such as Bacon and, after him, the Royal Society Fellows, to "usurp" Sidney's claim that poetry is the most effective science at moving humankind to "act virtuously."[50] Harrison might add that one of the primary ways the Royal Society Fellows advertised their usefulness to religion was to argue that experimental philosophy strengthened the moral values of its practitioners.[51] Simultaneously echoing Gosson and building on the Restoration's fears over divisive rhetoric, Sprat intimates that where poetry fails to instill virtue in its audience, experimental philosophy succeeds. Sprat argues that where God has historically "convince[d] the World from *Heaven* in a visible manner," such tactics "are not often now to be expected" (363). This lack of explicit divine guidance shores up Sprat's argument that:

> it is now the fittest season for *Experiments* to arise, to teach us a Wisdom, which springs from the depths of *Knowledge*, to shake off the shadows, and to scatter the mists, which fill the minds of men with a vain consternation. This is a *work* well-becoming the most *Christian Profession*. For the most apparent effect, which attended the passion of *Christ*, was the putting of an eternal silence on all the false oracles and dissembled inspirations of *Ancient Times*.
>
> (362–3)

This was an age, Sprat suggests, when experimental philosophy would call humankind to righteous belief. Having appropriated the poet's power to call an audience to action, Royal Society Fellows could more easily align their new experimental philosophy with theology.

Royal Society Fellows further worked to show how experimental philosophy buttressed theology by claiming that it offered evidence for God's existence and celebrated God's creations. Royal Society Fellows renewed the much older notion that God communicated primarily through the book of words in scripture and the book of works in nature. The powerful trope of the Book of Nature builds on Psalm 19:1, "The heavens declare the glory of God, the skies proclaim the work of his hands," and Romans 1:20, "For since the creation of the world God's invisible qualities—his eternal power and divine nature—have been clearly seen, being understood from what has been made."[52] Sprat harkens back to these biblical passages, selecting the Book of Nature as the domain that experimental philosophers are best equipped to study, urging experimental philosophers to

> meddle no otherwise with *Divine things*, than only as the *Power*, and *Wisdom*, and *Goodness* of the *Creator* is displayed in the admirable order and workmanship of the Creatures. It cannot be denied, but it lies in the *Natural Philosopher*'s hands, best to advance that part of *Divinity*.
>
> (82)[53]

By defining the Royal Society's domain of authority as the Book of Nature, he leaves the "divine Things," the book of words, to the theologians he seeks to appease. But Sprat's rhetorical move does not divorce experimental philosophy from theology; it merely suggests that they perform commensurable work in different domains. For Harrison, "experimental philosophy was widely acknowledged to provide internal evidence for God's existence, primarily by educing the many instances of design in nature" (*Territories*, 112). As Sprat maintains, studying the Book of Nature was the original way in which to know and worship God:

> This was the first service that *Adam* performed to his *Creator*, when he obeyed him [...] [by] looking into the *Nature* of all the *Creatures*. This had been the only *Religion*, if men had continued innocent in *Paradise* and had not wanted a *Redemption*.
>
> (349–50)

For Sprat, experimental philosophy was the original form of religious study, cementing the Royal Society's role as an aid to theologians. This was a conviction Robert Boyle also shared. For him, "discovering to others the

Perfections of God displayed in the Creatures is a more acceptable act of Religion than the burning of Sacrifices or Perfumes upon his Altars."[54] Boyle went so far as to consider the natural world a place of divine worship and to describe experimental philosophers as priests: "It will not be rash to infer that if the World be a Temple, Man sure must be the Priest, ordained [...] to celebrate Divine service not only in it, but for it."[55] Experimental philosophers aimed to discover in nature what they believed Adam had in paradise and, in the process, reveal the wondrous Edenic qualities of God's creations.

By revealing the wonders of nature, experimental philosophy for Sprat and science for Dawkins takes on a poetic quality. While Sidney's poet discovered a "golden" world that made nature's divine order visible, Sprat proclaims that the experiment reveals before ones "eyes the *beauty*, *contrivance*, and *order* of *God's Works*" (349). Sprat emphasizes the poetic quality of experimental philosophy because it bolstered the Royal Society's claims that "*Experiments* [are] *not dangerous to the Christian Religion*" (345) nor "*injurious to the worship of God*" (349). More than through the poet's pen, Sprat argues that experimental philosophy reveals the divine signatures that he thought God imprinted upon all of creation. The promise to reveal a "golden" world in nature itself situated the experimental philosopher both apart from the poet and alongside the theologian as an important aid to religious belief. The tactics Royal Society Fellows used to gain authority over poetry were the same tactics that they used to legitimate experimental philosophy within theology. The Royal Society thus brought the poetic quality of their experimental philosophy directly to bear on their relationship to theology. But where the early Royal Society Fellows appropriated poetry's power to align their experimental philosophy with religion, Dawkins appropriates the power of poetry to render religion obsolete.

Dawkins wants to evacuate religion from the domains of science, if not eliminate it entirely from society, but relies on the same arguments that Sprat used to ward off any accusations of heresy. For Dawkins, a poetic science cultivates a sense of wonder, making it a superior replacement to what he considers to be the irrational enchantments of religion. To borrow Atkins's turn of phrase, this notion of science's "omnicompetence" adopts and extends the seventeenth-century rhetoric of Sprat and early Royal Society Fellows. Although Sprat and his compatriots framed their new experimental philosophy in devout terms, the rhetoric of obsolescence that underpins their relationship to poetry remerges today in ways totally counter to the early Royal Society's aims. Harrison's assessment of science popularizers today thus rings true: "the otherwise puzzling antireligious rhetoric of some contemporary scientific popularizers is thus to be understood as a kind of inversion of a long-standing historical relationship that bound the study of nature to deeper moral and religious concerns" (*Territories*, 180). The means by which Dawkins delimits science from religion, then, are the very same Sprat used to usher experimental philosophy into theologians' good graces.

The same rhetorical thread about poetry wove its way throughout the centuries, stitching together a type of scientific authority that would consider a rainbow a testament to the religious utility of Sprat's experimental philosophy and an occasion to celebrate the obsolescence of religion to Dawkins's science. Sprat's aim to legitimize experimental philosophy as a complementary pursuit to theology relies on the rhetorical strategies that he and other Royal Society Fellows used to appropriate the work of the poet. First, Sprat argues for the moral utility of this new form of natural philosophy, rendering the poet's work moot in relation to experimental philosophy. Second, Sprat's reliance on the renewed trope of the Book of Nature presupposes the superiority of the experimental philosopher to discover the golden world once revealed by poets and signals a return to what Royal Society Fellows considered the original form of worship. Experimental philosophers leveraged their relationship to poetry to align their work with theology, knotting together all three fields of knowledge. Dawkins's recent attempt to bind science and literature together to diminish religion's authority replicates and inverts the tangled connections that founding Royal Society Fellows established hundreds of years prior. These knotted relationships demonstrate how scholars can move beyond conflict models while still accounting for the divergence among science, literature, and religion. The tension that conflict models ascribe to science, literature, and religion is best understood as a competition, not a "conflict," because their practitioners shared the same epistemological goal: to magnify and revere what they understood to be God's divine works. Indeed, experimental philosophers competed with poets for authority over wonder and imagination, mobilizing this poetic power to better align experimental philosophy with the work of the theologian. It is on the basis of these forms of appropriation that scholars can better understand how the boundaries and types of authority ascribed to literature, science, and religion change over time. The rhetorical and ideological threads that link Sprat to Dawkins attest to the persistent entanglement among science, literature, and religion. The "cold philosophy" of science cannot "unweave a rainbow" after all.

Notes

1 I am indebted to the careful attention and insights of Jeffrey Wigelsworth, Marshall Brown, and the anonymous reader at Bloomsbury for their comments on various drafts of this essay. Dawkins, *Unweaving the Rainbow*, 18. Hereafter cited parenthetically in the text as *Rainbow* followed by page number.
2 Atkins, "The Limitless Power of Science," 123.
3 Evelyn Fox Keller argues that she was "blackballed" from scientific communities for proposing to study the rhetoric of the Human Genome Project. See "Can the Circles of Language and Science be Squared?," 14.
4 Dawkins stipulates that his book title, *The God Delusion*, "does not refer to the God of Einstein and the other enlightened scientists [...] [he is instead] talking only about *supernatural* gods," including the Gods of major monotheistic religions (41). Further references to this work are hereafter cited parenthetically in the text as *Delusion* followed by page number.
5 Atkins, "Limitless Power of Science," 132, 123.
6 Albanese, *New Science, New World*, 3.
7 Harrison, "Introduction," 4.
8 Campbell, "Literature," 3:756.
9 Ibid., 3:757.
10 Marchitello, *The Machine in the Text*, 3.
11 Chico, *The Experimental Imagination*, 5.
12 Harrison, "Introduction," 4.
13 Osler, "Mixing Metaphors" 93. Osler observes that "while conflict has been the focus of historians celebrating the rise of science and its triumph over religion, harmony has often been the emphasis of religious apologists" ("Mixing Metaphors," 95). Hereafter cited parenthetically in the text by page number.
14 There is, of course, robust scholarship on the relationship between early modern literature and religion. Ken Jackson and Arthur F. Marotti address this field of scholarship at length in "The Turn to Religion in Early Modern English Studies," 167–90.
15 Smith laments that studies of the eighteenth century "have barely begun to register the extent to which literature and language were caught up in the same project that motivated eighteenth-century science and religion: the search for divine 'Truths'" (*Empiricist Devotions*, 2).
16 Reid Barbour's *Literature and Religious Culture in Seventeenth-Century England*, Joanna Picciotto's *Labors of Innocence in Early Modern England*, and Courtney Weiss Smith's *Empiricist Devotions* are notable exceptions.
17 Mazzio, "Shakespeare and Science," c. 1600," 1.
18 Park and Daston, "Introduction," *Cambridge History of Science*, 3:3.
19 Mazzio, "Shakespeare and Science," 1.

20 For this reason, I use the terms "natural philosophy," "experimental philosophy," "poetry," and "theology" when addressing the seventeenth century and the terms "science," "literature," and "religion" when discussing our current era.
21 Gosson, *The School of Abuse.* When quoting Gosson, I have opted to modernize only those aspects of the text that might occlude its meaning to modern readers.
22 Ibid., 47.
23 Sidney, "Defense," 236, 220. Hereafter cited parenthetically in the text by page number.
24 Heninger Jr., *Sidney and Spenser*, 293.
25 Osler usefully delineates "religion" from "theology," defining the former as "doctrine, faith, and practice" and the latter as the "enterprise of explaining the meaning of religious doctrines or practices" ("Mixing Metaphors," 92).
26 Cressy and Ferrell, *Religion and Society in Early Modern England*, 1.
27 Feldhay, "Religion," 3:731.
28 Harrison, *The Territories of Science and Religion*, 31. Hereafter cited parenthetically in the text as *Territories* followed by page number.
29 Gaukroger, *The Emergence of a Scientific Culture*, 23.
30 Butler, "The Elephant in the Moon in Long Verse," 17. It is important to note, however, that Butler did not reject the tenets of the new experimental philosophy, only the Royal Society's failure to adhere to them. As Ken Robinson reminds us, a growing body of scholarship demonstrates that Butler's poem "is a satire not on science but on bad scientists" ("The Skepticism of Butler's Satire on Science," 4).
31 Cavendish, *The Blazing World and Other Writings*, ed. Lilley, 142.
32 Swift, *Gulliver's Travels*, 151–52.
33 Even the Galileo Affair does not suggest an antagonism between religion and the new science. As Harrison contends, "in the case of Galileo, the Catholic Church was not opposing science per se. On the contrary, it was using its considerable authority to endorse what was then the consensus of the scientific community" ("Introduction," 5).
34 Robert South, quoted in Harrison, *Territories*, 125.
35 Keats, "Lamia," in *Selected Poems*, 217. Dawkins argues in *Unweaving the Rainbow* that Keats "could hardly have been more wrong" to have assumed that "Newton had destroyed all the poetry of the rainbow" (x). Mary Midgley admires Dawkins's decision to bridge poetry and science and "bring these two departments of life together," but explains that his demand that Romantic poets "draw […] topics from science" is an unfair ask (*Science and Poetry*, 78, 64).
36 Dawkins, *Science in the Soul*, 4.
37 Ibid., 80; Sagan, *Pale Blue Dot*, 50.
38 Sprat, *The History of the Royal-Society of London*. When quoting Sprat's *History*, I rely on the original edition published in 1667 and have opted to

modernize only those aspects of the text that might occlude its meaning to modern readers. Hereafter cited parenthetically in the text by page number.

39 The Royal Society's use of the very poetic language they rejected occasions much scholarly debate. Richard Nate enumerates a handful of Sprat's poetic ornaments, including Sprat's "delight in alliteration" and "his love for redundant enumerations," as well as potent allusions to prelapsarian Adam ("Quitting Now the Flowers of Rhetoric," 150). See also Richard Foster Jones, *The Seventeenth Century*; Brian Vickers, "The Royal Society and English Prose Style"; and Tina Skouen, "Science versus Rhetoric? Sprat's *History of the Royal Society* Reconsidered."

40 Hooke, *Micrographia*, Preface. When quoting Hooke's *Micrographia*, I have opted to modernize only those aspects of the text that might occlude its meaning to modern readers.

41 Skouen identifies the role of wonder in the rhetoric of other science popularizers, including Brian Cox and Paul Nurse, and notes the similarities between these current Royal Society Fellows and their seventeenth-century predecessors ("Brian Cox and the New Enlightenment").

42 Keats, "Lamia," in *Selected Poems*, 216, 217.

43 This argument is indebted to the perceptive insight of Jeffrey Wigelsworth. See also Sprat, *History*, 428, and Dawkins, *The God Delusion*, 319, 63.

44 Heyd, "The Reaction to Enthusiasm in the Seventeenth Century," 259.

45 Duran, "Reformed Catechism and Scientific Method," 80–1.

46 Bono, "Reading Literally," 181.

47 Smith, *Empiricist Devotions*, 3.

48 Funkenstein *Theology and the Scientific Imagination from the Middle Ages to the Seventeen Century*, 3. For Feldhay, this is a time when experimental philosophers and other naturalists "sacralize[d]" their knowledge by "couch[ing] their own claims in a new kind of religious rhetoric." But Feldhay also points out that experimental philosophers' recourse to theology was not a one-way street. She continues that "at the same time […] religious establishments, threatened by the fragmentation of Christianity, attempted to reclaim authority by connecting themselves to the new natural philosophy" ("Religion," 730).

49 Picciotto, *Labors of Innocence*; Harrison, *The Fall of Man and the Foundations of Science*, 4.

50 Gaukroger, *The Emergence of a Scientific Culture*, 199, 200.

51 Harrison, "The Fashioned Image of Poetry or the Regular Instruction of Philosophy?" 33.

52 All scriptural quotations taken from the New International Version.

53 Sprat elevates experimental philosophers yet again when he argues that "the greatness of the *Divine Majesty* is best to be worshipped by the due honoring and observing of *Nature*" (133).

54 Boyle, *Some Considerations Touching the Usefulness of Experimental Natural Philosophy*, Essay V, 115. When quoting Boyle, I have opted to modernize only those aspects of the text that might occlude its meaning to modern readers. See also Harrison, *Territories*, 111.

55 Boyle, Essay III, 55–6. See also Fisch, "The Scientist as Priest."

Bibliography

Ackerman, Robert. "The Debate of the Body and Soul in Parochial Christianity." *Speculum*, vol. 37, no. 4 (1962): 541–65.

Albanese, Denise. *New Science, New World*. Durham, NC: Duke University Press, 1996.

Anon. *Creation Museum Signs*. Green Forest, AK: Master Books, 2023.

Anon. "Dissection Viewed with Reference to the Resurrection." *The Boston Medical and Surgical Journal*, vol. 6, no. 13 (1832): 197–9.

Anon. *An Elegy on the Miraculously Learned and Much-Lamented Bishop of Armagh*. London, 1656.

Anon. *Journey Through the Creation Museum*. Green Forest, AK: Master Books, 2024.

Anon. "A Problem for Finding the Year of the Julian Period by a New and Very Easy Method." *Philosophical Transactions of the Royal Society*, vol. 1, no. 18 (1665): 324.

Anon. "Why Is Wednesday November 17, 1858, the Base Time for VAX/VMS?" Digital Equipment Corporation-Customer Support Center. Colorado Springs. June 6, 2007. https://web.archive.org/web/20070606064704/http://vms.tuwien.ac.at/info/humour/vms-base-time-origin.txt. Accessed December 9, 2024.

Answers in Genesis. "The Annals of the World, Written by James Ussher." https://answersingenesis.org/store/product/annals-world/?srsltid=AfmBOoo2e9xvfksQCS9qPiYzk4wfmheBOG7QKe1jybTteOh4zxb9-H7f. Accessed August 27, 2025.

Answers in Genesis. "Statement of Faith." https://answersingenesis.org/about/faith/. Accessed July 4, 2025.

Archimedes. *The Works of Archimedes*, ed. T. L. Heath. Mineola, NY: Dover Publications Inc., 2002.

Aristotle. *Metaphysics, Volume I: Books 1–9*, trans. Hugh Tredennick. Cambridge, MA: Harvard University Press, 1933.

Aristotle. *On the History of Animals*, trans. A. L. Peck. Cambridge, MA: Harvard University Press (Loeb Classical Library), 1965.

Ashworth, William B. "Emblematic Natural History of the Renaissance." In *Cultures of Natural History*, ed. Nicholas Jardine, J. A. Secord, and E. C. Spary. New York: Cambridge University Press, 1996: 17–37.

Atkins, Peter W. "The Limitless Power of Science." In *Nature's Imagination: The Frontiers of Scientific Vision*, ed. John Cornwell. Oxford: Oxford University Press, 1995: 122–32.

Attié, Katharine Bootle. "Tragic Proportions: The Art of Tyranny and the Politics and the Politics of the Soul in Hamlet." *English Literary Renaissance*, vol. 44, no. 1 (2014): 78–107.

Augustine. *Answer to Faustus, A Manichean*, trans. Roland Teske. New York: New City Press, 2007.

Augustine. *The City of God*, ed. and trans. Marcus Tods. New York: Modern Library, 1950.

Augustine. *City of God. Vol 1, Books 1–3*, trans. George E. McCracken. Cambridge, MA: Harvard University Press (Loeb Classical Library), 1957.

Augustine. *The City of God: Vol. IV, Books 12-15*, trans. Philip Levine. Loeb Classical Library 414. Cambridge, MA: Harvard University Press, 1966.

Augustine. *The City of God: Vol. VII, Books 21-22*, trans. William M. Green. Loeb Classical Library 417. Cambridge, MA: Harvard University Press, 1972.

Augustine. *Concerning the City of God Against the Pagans*, trans. Henry Bettenson. New York: Penguin Books, 1984.

Augustine. "Of the Catechizing of the Unlearned." In *Seventeen Short Treatises*, trans. C. L. Cornish and H. Brown. Oxford: John Henry Parker, 1847: 187–242.

Bacon, Francis. *The Advancement of Learning*, ed. W. Aldis Wright. Oxford: Clarendon Press, 1869.

Bacon, Francis. *The Advancement of Learning*, ed. Michael Kiernan. Oxford: Oxford University Press, 2000.

Bacon, Francis. *The Instauratio Magna Part II: Novum Organum and Associated Texts*, ed. Graham Rees and Maria Wakely. New York: Oxford University Press, 2004.

Bacon, Francis. "New Atlantis." In *Three Early Modern Utopias: Utopia, New Atlantis, and the Isle of Pines*, ed. Susan Bruce. Oxford World Classics. New York: Oxford University Press, 1999: 149–86.

Bacon, Francis. *Novum Organum*, ed. Lisa Jardine and Michael Silverthorne. Cambridge: Cambridge University Press, 2000.

Bacon, Francis. "Of Plantations." In *The Essayes or Counsels, Civill and Morall*, ed. Michael Kiernan. Cambridge, MA: Harvard University Press, 1985.

Bacon, Francis. *Works*, ed. James Spedding, Robert Leslie Ellis, and Douglas Denon Heath. 14 vols. London: Longman, 1857–74.

Bacon, Roger. *Roger Bacon and the Origins of Perspectiva in the Middle Ages: A Critical Edition and English Translation of Bacon's "Perspectiva" with Introduction and Notes*, ed. David C. Lindberg. Oxford: Oxford University Press, 1996.

Baker, Humfrey. *The Wellspring of Sciences: Teaching the Perfect Worke and Practice of Arithmetic*. London, 1646.

Bakhtin, Mikhail. *The Dialogic Imagination*, trans. Michael Holquist and Caryl Emerson. Austin: University of Texas Press, 1981.

Banister, John. *The History of Man*. London, 1578.

Barbour, Ian G. *When Science Meets Religion*. San Francisco, CA: HarperCollins, 2000.

Barbour, Reid. *Literature and Religious Culture in Seventeenth-Century England*. Cambridge: Cambridge University Press, 2002.

Barker, Francis. *The Tremulous Private Body: Essays on Subjection*. Ann Arbor: University of Michigan Press, 1995.

Barr, James. "Why the World Was Created in 4004 B.C.: Archbishop Ussher and Biblical Chronology." *Bulletin of the John Rylands University Library of Manchester*, vol. 67, no. 2 (1985): 575–608.

Baxter, Richard. *An Excellent Discourse*. London, 1682.

Baxter, Richard. *Methodus Theologiæ Christianæ*. London, 1681.

Beller, Sieghard and Andrea Bender. "The Limits of Counting: Numerical Cognition Between Evolution and Culture." *Science*, vol. 319, no. 5860 (2008): 213–15.

Bernard, Nicholas. *The Life & Death of the Most Reverend and Learned Father of Our Church Dr. James Ussher*. London, 1656.

Berry, Lloyd E. "Introduction." In *The Geneva Bible: A Facsimile of the 1560 Edition*, ed. Lloyd E. Berry. Peabody, MA: Hendrickson, 2007: 1–24.

Bessemer, Sir Henry. *An Autobiography*. London: Offices of Engineering, 1905.

La Bible qui est toute la saincte escriture, trans. Robert Olivétan. Neuchâtel: Pierre de Vingle, 1535.

Bickerman, E. J. *Chronology of the Ancient World*. Ithaca, NY: Cornell University Press, 1980.

Bicks, Caroline. *Midwiving Subjects in Shakespeare's England*. Burlington, VT: Ashgate, 2003.

Birken, William. "Drake, Roger (1608–1669), Physician and Minister of Religion." *Oxford Dictionary of National Biography*. September 23, 2004.

Blair, Ann. *The Theater of Nature: Jean Bodin and Renaissance Science*. Princeton, NJ: Princeton University, 1997.

Blair, Ann, and Kaspar von Greyerz. *Physico-Theology: Religion and Science in Europe, 1650–1750*. Baltimore, MD: Johns Hopkins University Press, 2020.

Blake, Liza. "After Life in Margaret Cavendish's Vitalist Posthumanism." *Criticism*, vol. 62, no. 3 (2020): 433–6.

Blank, Paula. *Shakespeare and the Mismeasure of Renaissance Man*. Ithaca, NY: Cornell University Press, 2006.

Blum, Paul Richard. "The Immortality of the Soul." In *The Cambridge Companion to Renaissance Philosophy*, ed. James Hankins. Cambridge: Cambridge University Press, 2007: 211–33.

Bly, Mary. *Queer Virgins and Virgin Queans on the Early Modern Stage*. Oxford: Oxford University Press, 2000.

Bodin, Jean. *Universae Naturae Theatrum*. Lyon, 1596.

Bond, David. *Negative Ecologies: Fossil Fuels and the Discovery of the Environment*. Minneapolis: University of Minnesota Press, 2022.

Bono, James J. "Reading Literally: Boyle, the Bible, and the Book of Nature." In *The Palgrave Handbook of Early Modern Literature and Science*, ed. Howard Marchitello and Evelyn Tribble. London: Palgrave Macmillan, 2017: 179–207.

Borlik, Todd Andrew, ed. *Literature and Nature in the English Renaissance: An Ecocritical Anthology*. Cambridge: Cambridge University Press, 2019.

Borst, Arno. *The Ordering of Time: From the Ancient Computus to the Modern Computer*, trans. Andrew Winnard. Cambridge: Polity Press, 1993.

Bossy, Michel-André. "Medieval Debates of Body and Soul." *Comparative Literature,* vol. 28, no. 2 (1976): 144–63.

Botkin, Daniel. *Discordant Harmonies: A New Ecology for the Twenty-First Century*. Oxford: Oxford University Press, 1992.

Bovon, François. "The Soul's Comeback: Immortality and Resurrection in Early Christianity." *The Harvard Theological Review*, vol. 103, no. 4 (2010): 387–406.

Bowerbank, Sylvia. *Speaking for Nature: Women and Ecologies of Early Modern England*. Baltimore, MD: Johns Hopkins University Press, 2004.

Boyce, Mark S. and Alan Haney, eds. *Ecosystem Management: Applications for Sustainable Forest and Wildlife Resources*. New Haven, CT: Yale University Press, 1997.

Boyle, Robert. *Some Considerations Touching the Usefulness of Experimental Natural Philosophy*, 2nd ed. Oxford, 1664.

Bradshaw, John. *A Concordance to the Poetical Works of John Milton*. London: George Allen and Unwin, 1965.

Brooke, John Headley. *Science and Religion: Some Historical Perspectives*. Cambridge: Cambridge University Press, 1991.

Brorby, Joshua. "Seeming as Believing: Epistemological Uncertainty and the World of Annus Mirabilis." *Restoration: Studies in English Literary Culture, 1660–1700*, vol. 43, no. 1 (2019): 29–50.

Brown, Vincent. *The Reaper's Garden: Death and Power in the World of Atlantic Slavery*. Cambridge, MA: Harvard University Press, 2008.

Browne, Sir Thomas. *The Works of Sir Thomas Browne*, ed. Geoffrey Keynes. 4 vols. London: Faber and Faber, 1964.

Bulhof, Ilse, and Laurens ten Kate. *Flight of the Gods: Philosophical Perspectives on Negative Theology*. New York: Fordham University Press, 2000.

Burke, Helen M. "Annus Mirabilis and the Ideology of the New Science." *ELH: English Literary History*, vol. 57, no. 2 (1990): 307–34.

Burton, Robert. *The Anatomy of Melancholy*, ed. Thomas C. Faulkner et al. 6 vols. Oxford: Clarendon Press, 1989–2000.

Burton, Robert. *The Anatomy of Melancholy*, ed. Holbrook Jackson and intro. William H. Gass. New York: New York Review of Books, 2001.

Bush, Douglas. *English Literature in the Earlier Seventeenth Century*. Oxford: Oxford University Press, 1973.

Butler, Samuel. "The Elephant in the Moon in Long Verse." In *Satires and Miscellaneous Poetry and Prose*, ed. René Lamar. Cambridge: Cambridge University Press, 1928: 17–30.

Bynum, Caroline Walker. *The Resurrection of the Body in Western Christianity, 200–1336*. New York: Columbia University Press, 1995.

Calloway, Katherine. *Literature and Natural Theology in Early Modern England*. Cambridge: Cambridge University Press, 2023.

Calvin, Jean. *Commentary on a Harmony of the Evangelists, Matthew, Mark, and Luke, Vol. 1*, trans. William Pringle. Edinburgh: Calvin Translation Society, 1845.

Campbell, Gordon. *Bible: The Story of the King James Version 1611 – 2011.* Oxford: Oxford University Press, 2011.
Campbell, Mary Baine. "Literature." In *The Cambridge History of Science: Early Modern Science, Vol. 3*, ed. Katharine Park and Lorraine Daston. 2006. Reprint Cambridge: Cambridge University Press, 2016: 756–72.
Cartari, Vincenzo. *Le imagini de i dei de gli antichi: nelle quali si contengono gl'idoli, riti, ceremonie, & altre cose appartenenti alla religione de gli antichi*. Venice, 1571.
Cavendish, Margaret. *The Blazing World and Other Writings*, ed. Kate Lilley. London: Penguin Books, 2004.
Cavendish, Margaret. *Grounds of Natural Philosophy, Divided into Thirteen Parts*. London, 1668.
Cavendish, Margaret. *Observations upon Experimental Philosophy*, ed. Eileen O'Neill. Cambridge: Cambridge University Press, 2001.
Cavendish, Margaret. *Observations upon Experimental Philosophy, To Which Is Added the Description of a New Blazing World*. London, 1666.
Cavendish, Margaret. *Paper Bodies: A Margaret Cavendish Reader*, ed. Sylvia Bowerbank and Sara Mendelson. New York: Broadview Press, 2000.
Cavendish, Margaret. *Philosophical Letters*. London, 1664.
Cellamare, Davide. "Anatomy and the Body in Renaissance Protestant Psychology." *Early Science and Medicine*, vol. 19, no. 4 (2014): 341–64.
Cheng, Mingna. "Master or Handmaid?: Dryden's View of the New Science in Annus Mirabilis." *The Seventeenth Century*, vol. 38, no. 4 (2023): 675–90.
Chico, Tita. *The Experimental Imagination: Literary Knowledge and Science in the British Enlightenment*. Stanford, CA: Stanford University Press, 2018.
Chovanec, Kevin. "Stirring Meats: Resurrection Doctrine and the Substance of the Body in the Revenger's Tragedy." *Religion & Literature*, vol. 48, no. 1 (2016): 1–25.
Clucas, Steven. "The Atomism of the Cavendish Circle: A Reappraisal." *The Seventeenth Century*, vol. 9, no. 2 (1994): 247–73.
Colenso, John William. *The Pentateuch and the Book of Joshua Critically Examined*. London: Longman, Green Longman, Roberts & Green, 1862.
Collingwood, R. G. *The Idea of Nature*. Oxford: The Clarendon Press, 1945.
Compier, Abdul Hoq."Rhazes in the Renaissance of Vesalius." *Medical History*, vol. 56, no. 1 (2012): 3–25.
Coolman, Boyd Taylor. "Twelfth Century Theology." In *The Oxford Handbook of Sacramental Theology*, ed. Hans Boersma and Matthew Levering. Oxford: Oxford University Press, 2015: 201–17.
Cooper, Amy. "Francis Bacon's Idols and the Reformed Science." *Studies in Philology*, vol. 116, no. 2 (2019): 328–50.
Copenhaver, Brian P., ed. *Hermetica: The Greek Corpus Hermeticum and the Latin Asclepius in a New English Translation, with Notes and Introduction*. New York: Cambridge University Press, 1992.
Corbett, Margery, and Ronald Lightbown. *The Comely Frontispiece: The Emblematic Title-Page in England 1550–1660*. London: Routledge & Kegan Paul, 1979.

Corneanu, Sorana. "The Nature and Care of the Whole Man: Francis Bacon and Some Late Renaissance Contexts." *Early Science and Medicine*, vol. 22, no. 2/3 (2017): 130–56.

Corneanu, Sorana. *Regimens of the Mind: Boyle, Locke, and the Early Modern "Cultura Animi" Tradition*. Chicago, IL: The University of Chicago Press, 2011.

Cowan, Jacqueline L. "The Imagination's Arts: Poetry and Natural Philosophy in Bacon and Shakespeare." *Studies in Philology*, vol. 113, no. 1 (2016): 132–62.

Coward, Harold, and Toby Foshay, eds. *Derrida and Negative Theology*. Albany: State University of New York Press, 1992.

Cowley, Abraham. *The Complete Works in Verse and Prose of Abraham Cowley, in Two Volumes*. New York: AMS Press, Inc., 1967.

Crane, Mary. *Losing Touch with Nature: Literature and the New Science in 16th-Century England*. Baltimore, MD: Johns Hopkins University Press, 2014.

Creation Museum Website. https://creationmuseum.org/. Accessed December 7, 2024.

Cressy, David, and Lori Anne Ferrell. *Religion and Society in Early Modern England: A Sourcebook*. 2nd ed. New York: Routledge, 2005.

Crooke, Helkiah. *Mikrokosmographia: A Description of the Body of Man*. London, 1615.

Crystal, David. *The Cambridge Encyclopedia of Language*. Cambridge: Cambridge University Press, 1987.

Cunningham, Andrew. *The Anatomical Renaissance: The Resurrection of the Anatomical Projects of the Ancients*. Aldershot: Scolar Press, 1997.

Cunningham, Andrew. "Protestant Anatomy." In *Religious Confessions and the Sciences in the Sixteenth Century*, ed. J. Helm and A. Winkelmann. Leiden, the Netherlands: Brill, 2001: 44–50.

Cunningham, Jack. *James Ussher and John Bramhall: The Theology and Politics of Two Irish Ecclesiastics of the Seventeenth Century*. London: Routledge, 2007.

Cushman, Helen. "Handling Knowledge: Holy Bodies in the Middle English Mystery Plays." *Journal of Medieval and Early Modern Studies*, vol. 47, no. 2 (2017): 279–304.

Da Vinci, Leonardo. *The Notebooks*, trans. and ed. Jean Paul Richter. 2 vols. New York: Dover Publications, 1970.

Da Vinci, Leonardo. *Thoughts on Art and Life*, trans. Maurice Baring. Boston, MA: Marrymount, 1906.

Dal Prete, Ivano. *On the Edge of Eternity: The Antiquity of the Earth in Medieval and Early Modern Europe*. Oxford: Oxford University Press, 2022.

Das, Nandini, and Nick Davis. "Introduction." In *Enchantment and Dis-enchantment in Shakespeare and Early Modern Drama: Wonder, the Sacred, and the Supernatural*, ed. Nandini Das and Nick Davis. London: Routledge, 2016: 1–17.

Daston, Lorraine J. "Marvelous Facts and Miraculous Evidence in Early Modern Europe." *Critical Inquiry*, vol. 18, no. 1 (1991): 93–124.
Davies, Abe. *Imagining the Soul in Premodern Literature*. London: Palgrave Macmillan, 2021.
Davies, John. *Nosce Teipsum*. London, 1599.
Dawkins, Richard. *The God Delusion*. New York: Houghton Mifflin, 2006; reprint, New York: First Mariner Books, 2008.
Dawkins, Richard. *Science in the Soul: Selected Writings of a Passionate Rationalist*, ed. Gillian Somerscales. New York: Random House, 2017.
Dawkins, Richard. *Unweaving the Rainbow: Science, Delusion, and the Appetite for Wonder*. Boston, MA, and New York: Houghton Mifflin, 1998; reprint, First Mariner Books, 2000.
De Laune, Thomas. *The Present State of London, Or, Memorials Comprehending a Full and Succinct Account of the Ancient and Modern State Thereof*. London, 1681.
DeCook, Travis. "The Ark and Immediate Revelation in Francis Bacon's New Atlantis." *Studies in Philology*, vol. 105, no. 1 (2008): 103–22.
Delbourgo, James. *Collecting the World: Hans Sloane and the Origins of the British Museum*. Cambridge, MA: Harvard University Press, 2019.
Delbourgo, James. "Sir Hans Sloane's Milk Chocolate and the Whole History of the Cacao." *Social Text*, vol. 29, no. 1 (2011): 71–101.
Descartes, René. *The Passions of the Soul*, trans. Stephen Voss. Indianapolis, IN: Hackett Publishing, 1989.
Descartes, René. *"The World" and Other Writings*, trans. and ed. Stephen Gaukroger. Cambridge: Cambridge University Press, 1998.
Diamond, Eli. *Mortal Imitations of Divine Life: The Nature of the Soul in Aristotle's "De anima."* Evanston, IL: Northwestern University Press, 2015.
Digby, Kenelm et al. *A Collection of Letters and Poems*. London, 1678.
Digby, Kenelm. *Observations upon Religio Medici*. London, 1643.
Digby, Kenelm. *Two Treatises*. Paris, 1644.
Dolan, Frances E. *True Relations: Reading, Literature, and Evidence in Seventeenth-Century England*. Philadelphia: University of Pennsylvania Press, 2013.
Donne, John. *The Complete English Poems*, ed. A. J. Smith. New York: Penguin, 1996.
Donne, John. *Complete Poems*, ed. Robin Robbins. London: Routledge, 2013.
Donne, John. *Devotions upon Emergent Occasions and Death's Duel with the Life of Dr. John Donne by Izaak Walton*. New York: Vintage Books, 1999.
Drake, Roger. *Sacred Chronology*. London, 1648.
Dryden, John. *Essays, Volume I & II*, ed. W. P. Ker. Oxford: Clarendon Press, 1900.
Dryden, John. *Works*, ed. H. T. Swedenberg Jr., Edward Niles Hooker, and Vinton A. Dearing. 20 vols. Berkeley: University of California Press, 1956–2002.
Dryden, John. *The Works of John Dryden, Now First Collected in Eighteen Volumes: vol. 9*, ed. Walter Scott. London, 1808.

Du Bartas, Guillaume. *Divine Weeks*, trans. Joshua Sylvester. London, 1605.

Du Bartas, Guillaume. *The Divine Weeks and Works of Guillaume de Saluste, Sieur Du Bartas*, trans. Joshua Sylvester, ed. Susan Snyder. 2 vols. Oxford: Clarendon Press, 1979.

Duran, Angelica. "Reformed Catechism and Scientific Method in Milton's of Education and Paradise Lost." In *Science, Literature and Rhetoric in Early Modern England*, ed. Juliet Cummins and David Burchell. Aldershot, Hampshire: Ashgate, 2007: 74–94.

Eagleton, Terry. *Reason, Faith, & Revolution: Reflections on the God Debate*. New Haven, CT: Yale University Press, 2009.

Eamon, William. *Science and the Secrets of Nature: Books of Secrets in Medieval and Early Modern Culture*. Princeton, NJ: Princeton University Press, 1996.

Ecklund, Elaine Howard, and Christopher P. Scheitle. *Religion vs. Science: What Religious People Really Think*. New York: Oxford University Press, 2018.

Eggert, Katherine. *Disknowledge: Literature, Alchemy, and the End of Humanism in Renaissance England*. Philadelphia: University of Pennsylvania Press, 2015.

Elyot, Thomas. *Dictionary*, 1538. Menston, England: Scholar Press Facsimile, 1970.

Empson, William. "Donne the Space Man." *Kenyon Review*, vol. 19, no. 3 (1957): 337–99.

Ettenhuber, Katrin. *The Logical Renaissance: Literature, Cognition, and Argument, 1479–1630*. Oxford: Oxford University Press, 2023.

Evelyn, John. *The Diary and Correspondence of John Evelyn, in Four Volumes*, ed. William Bray. London: George Bell & Sons, 1886–89.

Fauvel, John, and Jeremy Gray, eds. *The History of Mathematics: A Reader*. New York: Macmillan Press, 1987.

Feldhay, Rivka. "Religion." In *The Cambridge History of Science: Early Modern Science: Vol. 3*, ed. Katharine Park and Lorraine Daston. Cambridge: Cambridge University Press, 2016: 727–55.

Ferguson, Margaret. "Hymeneal Instruction." In *Masculinities, Childhood, Violence*, ed. Amy E. Leonard and Karen L. Nelson. Newark: University of Delaware Press, 2011: 97–130.

Finucci, Valeria. "Vesalius and the Languages of Anatomy." *Journal of Medieval and Early Modern Studies*, vol. 48, no. 1 (2018): 1–9.

Fisch, Harold. "The Scientist as Priest: A Note on Robert Boyle's Natural Theology." *Isis* 44, no. 3 (1953): 252–65.

Ford, Alan. "Ussher, James (1581–1656), Church of Ireland Archbishop of Armagh and Scholar." *Oxford Dictionary of National Biography*. September 23, 2004.

Foucault, Michel. *The History of Sexuality, Vol. 4: Confessions of the Flesh*, trans. Robert Hurley. New York: Vintage, 2021.

Freud, Sigmund. "The Taboo of Virginity" (1918). In *The Standard Edition of the Complete Psychological Works of Sigmund Freud*, ed and trans. James Strachey. 24 vols. London: The Hogarth Press, 1957: 11: 192–208.

Freudenthal, Gad. "Clandestine Stoic Concepts in Mechanical Philosophy: The Problem of Electrical Attraction." In *Renaissance and Reformation: Humanists, Scholars, Craftsmen, and Natural Philosophers in Early Modern Europe*, ed. J. V. Field and Frank A. J. L. James. Cambridge: Cambridge University Press, 1993: 107–16.

Freudenthal, Gad. "The Problem of Cohesion Between Alchemy and Natural Philosophy: From Unctuous Moisture to Phlogiston." In *Alchemy Revisited: Proceedings of the International Conference on the History of Alchemy at the University of Groningen*, ed. Z. R. W. M. von Martels. Leiden, the Netherlands: Brill, 1990: 107–16.

Fuller, J. G. C. M. "A Date to Remember: 4004 BC." *Earth Sciences History*, vol. 24, no. 1 (2005): 5–14.

Funkenstein, Amos. *Theology and the Scientific Imagination from the Middle Ages to the Seventeen Century*. Princeton, NJ: Princeton University Press, 1986.

Galen. *On the Usefulness of the Parts of the Body (De usu partium)*, trans. Margaret Tallmadge May. 2 vols. Ithaca, NY: Cornell University Press, 1968.

Gardner, Martin. "Mathematical Games." *Scientific American*, vol. 237, no. 5 (1977): 18–29.

Gaukroger, Stephen. "The Autonomy of Natural Philosophy." In *The Science of Nature in the Seventeenth Century: Patterns of Change in Early Modern Natural Philosophy*, ed. Peter R. Anstey and John A. Schuster. Dordrecht: Springer, 2005: 131–63.

Gaukroger, Stephen. *The Emergence of a Scientific Culture: Science and the Shaping of Modernity, 1210–1685*. Oxford: Oxford University Press, 2006; 2013.

Giglioni, Guido. "Francis Bacon." In *The Oxford Handbook of British Philosophy in the Seventeenth Century*, ed. Peter R. Anstey. Oxford: Oxford University Press, 2013: 41–72.

Gil, Daniel Juan. *Fate of the Flesh: Secularization and Resurrection in the Seventeenth Century*. New York: Fordham University Press, 2021.

Glanvill, Joseph. *The Vanity of Dogmatizing*. London, 1661.

Glimp, David. "Paradisal Arithmetic: Paradise Lost and the Genesis of Populations." *Modern Language Quarterly*, vol. 60, no. 1 (1999): 1–31.

Godwin, Joscelyn. *The Pagan Dream of the Renaissance*. Grand Rapids, MI: Phanes Press, 2002.

Goldhill, Simon. *Foucault's Virginity: Ancient Erotic Fiction and the History of Sexuality*. Cambridge: Cambridge University Press, 1995.

Gosson, Stephen. *The School of Abuse*. London, 1587.

Gould, Stephen Jay. *Eight Little Piggies: Reflections in Natural History*. New York: W. W. Norton, 1993.

Govier, Mark. "The Royal Society, Slavery, and the Island of Jamaica: 1660–1700." *Notes and Records of the Royal Society of London*, vol. 53, no. 2 (1999): 203–17.

Grafton, Anthony. "Joseph Scaliger and Historical Chronology: The Rise and Fall of a Discipline." *History and Theory*, vol. 14, no. 2 (1975): 156–85.
Grafton, Anthony. *Worlds Made by Words: Scholarship and Community in the Modern West*. Cambridge, MA: Harvard University Press, 2009.
Graham, Kenneth J. E., and Philip D. Collington, eds. *Shakespeare and Religious Change*. London: Palgrave, 2009.
Grant, Edward. *Science and Religion, 400 B.C. to A.D. 1500: From Aristotle to Copernicus*. Baltimore, MD: Johns Hopkins University Press, 2004.
Green, S. G. "Bases of Scripture Chronology." In *The King's Printer's Aids to the Student of the Holy Bible*, ed. C. J. Ball. London: Eyre and Spottiswoode, 1898: 97–104.
Green, William Henry. *The Pentateuch Vindicated from the Aspersions of Bishop Colenso*. New York: John Wiley, 1863.
Greenblatt, Stephen. *Hamlet in Purgatory*. Princeton, NJ: Princeton University Press, 2001.
Greenblatt, Stephen. "The Mousetrap." In *Practicing New Historicism*, ed. Greenblatt and Catherine Gallagher. Chicago, IL: University of Chicago, 2000: 232–42.
Greenblatt, Stephen. *Renaissance Self-Fashioning from More to Shakespeare*. Chicago, IL: University of Chicago Press, 1980.
Gregory, Brad. *The Unintended Reformation*. New York: Cambridge University Press, 2012.
Grosholz, Emily. "Mathematics." In *Encyclopedia of the Scientific Revolution: From Copernicus to Newton*, ed. Wilbur Applebaum. New York: Garland, 2000: 394–8.
Gruber, Elizabeth D. *The Eco-Self in Early Modern English Literature*. Amsterdam: Amsterdam University Press, 2023.
Guibbory, Achsah. "Imitation and Originality: Cowley and Bacon's Vision of Progress." *Studies in English Literature*, vol. 29, no. 1 (1989): 99–120.
Guilmartin, John. *Galleons and Galleys*. London: Cassell, 2002.
Haekel, Ralf. *The Soul in British Romanticism: Negotiating Human Nature in Philosophy, Science, and Poetry*. Trier, Germany: WVT, 2014.
Hallam, Anthony. *Great Geological Controversies*. Oxford: Oxford University Press, 1989.
Hannah, Robert. *Time in Antiquity*. London: Routledge, 2009.
Harding, Sandra. *The Science Question in Feminism*. Ithaca, NY: Cornell University Press, 1986.
Harkness, Deborah E. *The Jewel House: Elizabethan London and the Scientific Revolution*. New Haven, CT: Yale University Press, 2007.
Harrison, Peter. *The Bible, Protestantism, and the Rise of Natural Science*. Cambridge: Cambridge University Press, 1997.
Harrison, Peter. "The Book of Nature and Early Modern Science." In *The Book of Nature in Early Modern and Modern History*, ed. Klaas van Berkel and Arjo Vanderjagt. Leuven, Belgium: Peeters, 2006: 1–26.

Harrison, Peter. *The Fall of Man and the Foundations of Science*. Cambridge: Cambridge University Press, 2001.

Harrison, Peter. "'The Fashioned Image of Poetry or the Regular Instruction of Philosophy': Truth, Utility, and the Natural Sciences in Early Modern England." In *Science, Literature and Rhetoric in Early Modern England*, ed. Juliet Cummins and David Burchell. New York: Routledge, 2007: 14–35.

Harrison, Peter. "Francis Bacon, Natural Philosophy, and the Cultivation of the Mind." *Perspectives on Science*, vol. 20, no. 2 (2012): 139–58.

Harrison, Peter. "Introduction." In *The Cambridge Companion to Science and Religion*, ed. Harrison. Cambridge: Cambridge University Press, 2010: 1–17.

Harrison, Peter. "Science and Secularization." *Intellectual History Review*, vol. 27, no. 1 (2017): 47–70.

Harrison, Peter. *The Territories of Science and Religion*. Chicago, IL: The University of Chicago Press, 2015.

Harvey, Elizabeth D., ed. *Sensible Flesh: On Touch in Early Modern Culture*, Philadelphia: University of Pennsylvania Press, 2003.

Harvey, Elizabeth, and Timothy Harrison. *John Donne's Physics*. Chicago, IL: University of Chicago Press, 2024.

Harvey, William. *The Anatomical Exercises "De Motu Cordis" and "De Circulatione Sanguinis" in English Translation*, ed. Geoffrey Keynes. New York: Dover, 1995.

Hazlitt, William. *Characters of Shakespeare's Plays*. 3rd ed. London, 1838.

Helm, Jürgen. "Religion and Medicine: Anatomical Education and Wittenberg and Ingolstadt." In *Religious Confessions and the Sciences in the Sixteenth Century*, ed. J. Helm and A. Winkelmann. Leiden, the Netherlands: Brill, 2001: 51–68.

Heninger Jr., S. K. *Sidney and Spenser: The Poet as Maker*. University Park: Pennsylvania State University Press, 1989.

Henry, John. "The Matter of Souls: Medical Theory and Theology in Seventeenth-Century England." In *The Medical Revolution of the Seventeenth Century*, ed. Roger French and Andrew Wear. Cambridge: Cambridge University Press, 1989: 87–113.

Heseler, Baldasar. *Andreas Vesalius' First Public Anatomy at Bologna 1640*, ed. and trans. Ruben Eriksson. Upsala, Sweden: Almquist & Wiksells Boktryckeri AB, 1959.

Heyd, Michael. "The Reaction to Enthusiasm in the Seventeenth Century: Towards an Integrative Approach." *The Journal of Modern History*, vol. 53, no. 2 (1981): 258–80.

Hillman, David. "Visceral Knowledge: Shakespeare, Skepticism, and the Interior of the Early Modern Body." In *The Body in Parts: Fantasies of Corporeality in Early Modern Europe*, ed. David Hillman and Carla Mazzio. New York: Routledge, 1997: 81–105.

Hillman, David, and Carla Mazzio, eds. *The Body in Parts: Fantasies of Corporeality in Early Modern Europe*. London: Routledge 1997.

Hillyer, Richard. *Four Augustan Science Poets: Abraham Cowley, James Thomson, Henry Brooke, Erasmus Darwin*. London: Anthem Press, 2020.
Hinman, Robert. "'Truth Is Truest Poesy': The Influence of the New Philosophy on Abraham Cowley." *ELH: English Literary History*, vol. 23, no. 3 (1956): 194–203.
Hinman, Robert. *Abraham Cowley's World of Order*. Cambridge, MA: Harvard University Press, 1960.
Hitchens, Christopher. *God Is Not Great*. London: Atlantic, 2007.
Hobbes, Thomas. *The Metaphysical System of Hobbes*, ed. M. W. Calkins. Chicago, IL: The Open Court Publishing, 1910.
Hodges, Devon L. *Renaissance Fictions of Anatomy*. Amherst: University of Massachusetts Press, 1985.
Hooke, Robert. *Micrographia*. London, 1665.
Hunter, Michael. *Science and Society in Restoration England*. Cambridge: Cambridge University Press, 1981.
Hunter, Michael. "The Social Basis and Changing Fortunes of an Early Scientific Institution: An Analysis of the Membership of the Royal Society, 1660–1685." *Notes and Records of the Royal Society of London*, vol. 31, no. 1 (1976): 9–114.
Hutton, Clark. "Images of Time and Eternity in the Religious Poetry of Henry Vaughan." *Comitatus: A Journal of Medieval and Renaissance Studies*, vol. 36, no. 1 (2005): 142–56.
Iannini, Christopher. *Fatal Revolutions: Natural History, West Indian Slavery, and the Routes of American Literature*. Chapel Hill: University of North Carolina Press, 2012.
Jackson, Ken, and Arthur F. Marotti. "The Turn to Religion in Early Modern English Studies." *Criticism*, vol. 46, no. 1 (2004): 167–90.
Jennings, Gretchen. "Creationist 'Museums' Are Not Museums." *The Exhibitionist*, vol. 20 (2011): 72–4.
Johnson, John. *Johnson's Arithmetic in Two Books*. London, 1646.
Johnson, Samuel. *Lives of the English Poets, Volume I*. London: Oxford University Press, 1906.
Jones, Richard Foster. *Ancients and Moderns: A Study of the Rise of the Scientific Movement in Seventeenth-Century England*. St. Louis: Washington University Press, 1936; reissued 1961.
Jones, Richard Foster. *The Seventeenth Century: Studies in the History of English Thought and Literature from Bacon to Pope*. Stanford, CA: Stanford University Press, 1965.
Josephson-Storm, Jason Ā. "Max Weber and the Rationalization of Magic." In *Narratives of Disenchantment and Secularization: Critiquing Max Weber's Idea of Modernity*, ed. Robert A. Yelle and Lutz Trein. London: Bloomsbury Academic, 2021: 31–50.
Julian Day Calculator. https://www.tondering.dk/claus/cal/julperiod.php. Accessed December 7, 2024.

Kargon, Robert H. *Atomism in England from Hariot to Newton*. Oxford: Clarendon Press, 1966.

Keats, John. *Poetical Works*, ed. H. W. Garrod. Oxford: Oxford University Press, 1962.

Keats, John. *Selected Poems*, ed. John Barnard. London: Penguin Books, 2007.

Keller, Catherine. *Cloud of the Impossible: Negative Theology and Planetary Entanglement*. New York: Columbia University Press, 2014.

Keller, Evelyn Fox. "Can the Circles of Language and Science Be Squared?" In *Storytelling in Science and Literature*, ed. Margery Arent Safir. Lewisburg, WV: Bucknell University Press, 2015: 13–22.

Keller, Evelyn Fox. *Reflections on Gender and Science*. New Haven, CT: Yale University Press, 1985.

Kelly, Kathleen Coyne. *Performing Virginity and Testing Chastity in the Middle Ages*. London: Routledge, 2000.

Killeen, Kevin. *The Unknowable in Early Modern Thought: Natural Philosophy and the Poetics of the Ineffable*. Stanford, CA: Stanford University Press, 2013.

Kitch, Aaron. *Shakespeare's Theater of Nature: Science, Religion, and the Orders of Mimesis in Early Modern Europe*. London: Palgrave, 2025.

Kitch, Aaron. "Some Versions of Enchanted Materialism: Bennett, Paracelsus, Hobbes, and *Hamlet*." In *Enchantments and Dis-Enchantments in Early Modern England*, ed. Nick Davis and Nandini Das. London: Routledge, 2017: 72–87.

Kriz, Kay Dian. *Slavery, Sugar, and the Culture of Refinement: Picturing the British West Indies, 1700–1840*. New Haven, CT: Yale University Press, 2008.

Kuhn, Marcus. "A Summary of the International Standard Date and Time Notation." Available at https://fits.gsfc.nasa.gov/iso-time.html. Accessed December 7, 2024.

Kuhn, Thomas. "Mathematical vs. Experimental Traditions in the Development of Physical Science." *The Journal of Interdisciplinary History*, vol. 7, no. 1 (1976): 1–31.

Kuhn, Thomas. *The Structure of Scientific Revolutions*. Chicago, IL: University of Chicago Press, 1969.

Lagerlund, Henrik. "The Terminological and Conceptual Roots of Representation in the Soul in Late Ancient and Medieval Philosophy." In *Representation and Objects of Thought in Medieval Philosophy*, ed. Lagerlund. Farnham, Surrey: Ashgate, 2008: 11–32.

Lake, Peter. *Hamlet's Choice: Religion and Resistance in Shakespeare's Revenge Tragedies*. New Haven, CT: Yale University Press, 2020.

Landau, Iddo. "Feminist Criticisms of Metaphors in Bacon's Philosophy of Science." *Philosophy*, vol. 73, no. 1 (1998): 47–61.

Latour, Bruno. *Rejoicings, or the Torments of Religious Speech*, trans. Julie Rose. Cambridge: Polity, 2013.

Latour, Bruno. *We Have Never Been Modern*, trans. Catherine Porter. Cambridge, MA: Harvard University Press, 1993.

Laurens, André du (Laurentius). *L'Histoire anatomique*, trans. François Size. Paris, 1610.

Lee, John. *Shakespeare's "Hamlet" and the Controversies of the Self.* Oxford: Clarendon, 2000.

Lewis, Cherry. L. E. and Simon J. Knell, eds. *The Age of the Earth: From 4004 BC to AD 2002*. Bath: Geological Society, 2001.

Lewis, Rhodri. *Hamlet and the Vision of Darkness*. Princeton, NJ: Princeton University Press, 2017.

Liska, Anthony J. *Aquinas's Theory of Perception: An Analytic Reconstruction*. Oxford: Oxford University Press, 2016.

Lithgow, William. *The Present Survey of London and England's State*. London, 1643.

Locke, John. *An Essay Concerning Human Understanding*, ed. Peter H. Nidditch. Oxford: Clarendon Press, 1979.

Loughlin, Marie H. *Hymeneutics: Interpreting Virginity on the Early Modern Stage*. London: Associated University Presses, 1997.

Low, Anthony. "New Science and the Georgic Revolution in Seventeenth-Century English Literature." *English Literary Renaissance*, vol. 13, no. 3 (1983): 231–59.

MacGregor, Arthur, ed. *Sir Hans Sloane: Collector, Scientist, Antiquary, Founding Father of the British Museum*. London: British Museum Press, 1994.

Maifreda, Germano. *From Oikonomia to Political Economy: Constructing Economic Knowledge from the Renaissance to the Scientific Revolution*. Burlington, VT: Ashgate, 2012.

Mandressi, Rafael. "Of the Eye and the Hand: Performance in Early Modern Anatomy." *The Drama Review*, vol. 59, no. 3 (2015): 60–76.

Maor, Eli. *To Infinity and Beyond: A Cultural History of the Infinite*. Princeton, NJ: Princeton University Press, 1991.

Marchitello, Howard. *The Machine in the Text: Science and Literature in the Age of Shakespeare and Galileo*. Oxford: Oxford University Press, 2011.

Marchitello, Howard and Evelyn Tribble, eds. *The Palgrave Handbook of Early Modern Literature and Science*. London: Palgrave, 2017.

Margócsy, Dániel, Mark Somos, and Stephen N. Joffe. "The Art of Medicine: Vesalius Annotations and the Rise of Early Modern Medicine." *The Lancet*, vol. 393, no. 10173 (February 23, 2019): 738–9.

Margócsy, Dániel, Mark Somos, and Stephen N. Joffe. *The "Fabrica" of Andreas Vesalius: A Worldwide Descriptive Census, Ownership, and Annotations of the 1543 and 1555 Editions*. Leiden, Netherlands: Brill, 2018.

Marlowe, Christopher. *The Collected Poems of Christopher Marlowe*, ed. Patrick Cheney and Brian J. Striar. Oxford: Oxford University Press, 2006.

Marlowe, Christopher. *The Complete Works of Christopher Marlowe, Volume V: "Tamburlaine the Great," Parts 1 and 2; and "The Massacre at Paris,"* ed. David Fuller and Edward J. Esche. Oxford: Clarendon, 1998.

Marlowe, Christopher. *Dr. Faustus*, ed. David Scott Kastan. New York: Norton, 2005.

Martin, Raymond, and John Barresi. *The Rise and Fall of the Soul and Self: An Intellectual History of Personal Identity*. New York: Columbia University Press, 2006.

Marvell, Andrew. *The Poems of Andrew Marvell*, ed. Nigel Smith. New York: Longman, 2007.

Mason, Stephen F. *A History of the Sciences*. New York: Collier Books, 1962.

Masten, Jeffrey. *Queer Philologies: Sex, Language, and Affect in Shakespeare's Time*. Philadelphia: University of Pennsylvania Press, 2016.

Matthews, Steven. *Theology and Science in the Thought of Francis Bacon*. Aldershot, Hampshire: Ashgate, 2008.

Maus, Katharine. *Inwardness and Theater in the English Renaissance*. Chicago, IL: University of Chicago Press, 1995.

Mazzio, Carla. "Shakespeare and Science, c. 1600." *South Central Review*, vol. 26, no. 1 & 2 (Spring & Summer 2009): 1–23.

McCarthy, Daniel P. "The Biblical Chronology of James Ussher." *Irish Astronomical Journal*, vol. 24, no. 1 (1997): 73–82.

McClure, Peter, and Robin Headlam Wells. "Elizabeth I as a Second Virgin Mary." *Renaissance Studies*, vol. 4, no. 1 (1990): 38–70.

McColley, Diane Kelsey. *Poetry and Ecology in the Age of Milton and Marvell*. Farnham, Surrey: Ashgate, 2007.

McCoy, Marina Berzins. *Image and Argument in Plato's Republic*. Albany: State University of New York, 2020.

McGrath, Seth. *Thinking Nature: An Essay in Negative Ecology*. Edinburgh: Edinburgh University Press, 2019.

McKeon, Michael. *Politics and Poetry in Restoration England: The Case of Dryden's "Annus Mirabilis."* Cambridge, MA: Harvard University Press, 1975.

Mendelson, Sara. "The God of Nature and the Nature of God." In *God and Nature in the Thought of Margaret Cavendish*, ed. Brandie R. Siegfried and Lisa T. Sarasohn. London: Routledge, 2014: 31–49.

Merchant, Carolyn. *The Death of Nature: Women, Ecology and the Scientific Revolution*. San Francisco, CA: Harper & Row, 1980.

Merton, Robert King. *Science, Technology & Society in Seventeenth Century England*. New York: H. Fertig, 1970.

Midgley, Mary. *Science and Poetry*. London: Routledge, 2001.

Milton, John. *Paradise Lost*, ed. Alistair Fowler. 2nd ed. London: Routledge, 2013.

Mitchell, Peter. *The Purple Island and Anatomy in Early Seventeenth-Century Literature, Philosophy, and Theology*. Madison, NJ: Fairleigh Dickinson University Press, 2007.

Morland, Samuel. *A New Method of Cryptography*. London, 1666.

Morland, Samuel. *The Urim of Conscience*. London, 1695.

Morton, Timothy. *Dark Ecology: For a Logic of Future Coexistence*. New York: Columbia University Press, 2016.

Morton, Timothy. *The Ecological Thought*. Cambridge, MA: Harvard University Press, 2010.

Morton, Timothy. *Ecology Without Nature: Rethinking Environmental Aesthetics*. Cambridge, MA: Harvard University Press, 2007.

Morton, Timothy. *Hell: In Search of a Christian Ecology*. New York: Columbia University Press, 2024.

Morton, Timothy. *Hyperobjects: Philosophy and Ecology after the End of the World*. Minneapolis: University of Minnesota Press, 2013.

Moshenska, Joe. *Feeling Pleasures: The Sense of Touch in Renaissance England*. Oxford: Oxford University Press, 2014.

Moshenska, Joe. *A Stain in the Blood: The Remarkable Voyage of Sir Kenelm Digby*. London: William Heinemann, 2016.

Most, Glenn W. *Doubting Thomas*. Cambridge, MA: Harvard University Press, 2005.

Nate, Richard. "'Quitting Now the Flowers of Rhetoric': Anti-rhetorical Continuities in English Science and Literature." In *English Literature and the Disciplines of Knowledge, Early Modern to Eighteenth Century: A Trade for Light*, ed. Jorge Bastos da Silva and Miguel Ramalhete Gomes. Leiden, the Netherlands: Brill, 2018: 146–64.

Nethercot, Arthur H. *Abraham Cowley: The Muse's Hannibal*. New York: Russell & Russell, 1967.

Newman, Harry. "'A Seale of Virgin Waxe at Hand / Without Impression There Doeth Stand': Hymenal Seals in English Renaissance Literature." *Lives and Letters*, vol. 4, no. 1 (2012): 94–113.

Newman, Karen. *Fashioning Femininity and English Renaissance Drama*. Chicago, IL: University of Chicago Press, 1991.

North, J. D. "Chronology and the Age of the World." In *Cosmology, History, and Theology*, ed. Wolfgang Yourgrau and Allen D. Breck. New York: Plenum Press, 1977: 307–33.

Numbers, Ronald L. *Galileo Goes to Jail and Other Myths About Science and Religion*. Cambridge, MA: Harvard University Press, 2010.

Numbers, Ronald L. "'The Most Important Biblical Discovery of Our Time': William Henry Green and the Demise of Ussher's Chronology." *Church History*, vol. 69, no. 2 (2000): 257–76.

Nunn, Hillary M. *Staging Anatomies: Dissection and Spectacle in Early Stuart Tragedy*. Aldershot, Hampshire: Ashgate, 2005.

O'Malley, Charles Daniels. *Andreas Vesalius of Brussels, 1514–1564*. Berkeley: University of California Press, 1965.

O'Malley, Charles Daniels. "Helkiah Crooke." *Bulletin of the History of Medicine*, vol. 42, no. 1 (1967): 1–18.

Olson, Richard G. *Science and Religion, 1450–1900: From Copernicus to Darwin*. Westport, CT: Greenwood Press, 2004.

Osler, Margaret J. "Mixing Metaphors: Science and Religion or Natural Philosophy and Theology in Early Modern Europe." *History of Science*, vol. 35, no. 1 (1997): 91–113.

Otis, Jessica Marie. *By the Numbers: Numeracy, Religion, and the Quantitative Transformation of Early Modern England*. Oxford: Oxford University Press, 2024.

Ovid (Publius Ovidius Naso). *Metamorphoses*, ed. and trans. Frank Justus Miller. Rev. by G. P. Goold. 3d. ed. 2 vols. Loeb Classical Library 42 and 43. Cambridge, MA: Harvard University Press, 1977.

Ovid (Publius Ovidius Naso). *Metamorphoses*, trans. David Raeburn. New York: Penguin Books, 2004.

Park, Katharine. "Bacon's 'Enchanted Glass.'" *Isis*, vol. 75, no. 2 (1984): 290–302.

Park, Katharine. "The Organic Soul." In *The Cambridge History of Renaissance Philosophy*, ed. Charles B. Schmitt, Eckhard Kessler, and Jill Kraye. Cambridge: Cambridge University Press, 1988: 464–84.

Park, Katharine. *Secrets of Women: Gender, Generation, and the Origins of Human Dissection*. New York: Zone Books, 2006.

Park, Katharine, and Lorraine Daston. "Introduction: The Age of the New." In *The Cambridge History of Science: Early Modern Science, Vol. 3*, ed. Park and Daston. Cambridge: Cambridge University Press, 2006; 2016: 1–17.

Parr, Richard. *The Life of the Most Reverend Father in God, James Ussher, Late Lord Archbishop of Armagh*. London, 1686.

Paster, Gail Kern. *The Body Embarrassed: Drama and the Disciplines of Shame in Early Modern England*. Ithaca, NY: Cornell University Press, 1993.

Pastorino, Cesare. "The Philosopher and the Craftsman: Francis Bacon's Notion of Experiment and Its Debts to Early Stuart Inventors." *Isis*, vol. 108, no. 4 (2017): 749–68.

Pearson, Karl. *The History of Statistics in the 17th & 18th Centuries*. London: Charles Griffin, 1978.

Peltonen, Markku, ed. *The Cambridge Companion to Bacon*. New York: Cambridge University Press, 1996.

Pender, Steven. "Essaying the Body: Donne, Affliction, and Medicine." In *John Donne's Professional Lives*, ed. David Colclough. Cambridge, England: D. S. Brewer, 2003: 215–48.

Pérez-Ramos, Antonio. *Francis Bacon's Idea of Science and the Maker's Knowledge Tradition*. Oxford: Oxford University Press, 1988.

Peterson, Kaara L. "Elizabeth I's Virginity and the Body of Evidence: Jonson's Notorious Crux." *Renaissance Quarterly*, vol. 68, no. 3 (2015): 840–71.

Peterson, Robert. *Sir Kenelm Digby: The Ornament of England, 1603–1665*. Cambridge, MA: Harvard University Press, 1956.

Pfeffer, Jess R. "Trans Materiality: Crooke's *Mikrokosmographia*, Sexual Dimorphism, and the Embodiment of Identity." *Journal for Early Modern Cultural Studies*, vol. 19, no. 4 (2019): 227–41.

Phillip, Alexander. *The Calendar: Its History, Structure and Improvement*. Cambridge: Cambridge University Press, 1921.

Phillip, I. G. "A Seventeenth-Century Agreement Between Author and Printer." *Bodleian Library Record*, vol. 10, no. 1 (1978): 68–73.

Picciotto, Joanna. *Labors of Innocence in Early Modern England*. Cambridge, MA: Harvard University Press, 2010.

Pindar. *Nemean Odes, Isthmean Odes, Fragments*, ed. and trans William H. Race. Loeb Classical Library 485. Cambridge, MA: Harvard University Press, 1997.

Pitkin, Barbara. "The Reformation of Doubt: Incredulity, Faith, the Body, and the Search for Certitude in Early Lutheran Interpretations of John 20:24–31." In *Reading Certainty: Epistemology and Exegesis on the Threshold of Modernity*, ed. Ralph Keen, Elizabeth Palmer, and Daniel Owings. Leiden, the Netherlands: Brill, 2023: 136–74.

Pliny the Elder. *Natural History*, ed. and trans. H. Rackham, W. H. S. Jones, and D. E. Eichholz. 10 vols. Loeb Classical Library 330–419. Cambridge, MA: Harvard University Press, 1938–1962.

Pointon, Marcia R. *Strategies for Showing: Women, Possession, and Representation in English Visual Culture, 1665–1800*. New York: Oxford University Press, 1997.

Poole, Kristen. "God's Game of Hide-and-Seek: Bacon and Allegory." In *The Palgrave Handbook of Early Modern Literature and Science*, ed. Howard Marchitello and Evelyn Tribble. London: Palgrave Macmillan, 2017: 115–38.

Poole, Kristen. *Supernatural Environments in Shakespeare's England: Spaces of Demonism, Divinity, and Drama*. Cambridge, MA: Cambridge University Press, 2014.

Porter, Stephen, and Simon Marsh. *The Battle for London*. Stroud, Gloucestershire: Amberley Publishing, 2011.

Powell, Rosalind. *Perception and Analogy: Poetry, Science and Religion in the Eighteenth Century*. Manchester: Manchester University Press, 2021.

Powers, Amanda. *Roger Bacon and the Defense of Christendom*. Cambridge: Cambridge University Press, 2013.

Principe, Lawrence. *The Scientific Revolution: A Very Short Introduction*. Oxford: Oxford University Press, 2011.

Purchas, Samuel. *Purchas His Pilgrimage*. London, 1613.

Puttenham, George. *The Art of English Poesy by George Puttenham: A Critical Edition*, ed. Frank Whigham and Wayne A. Rebhorn. Ithaca, NY: Cornell University Press, 2007.

Quammen, David. *The Tangled Tree: A Radical New History of Life*. New York: Simon & Schuster, 2018.

Quarles, John. *An Elegy on the Most Reverend & Learned James Ussher*. London, 1656.

Raleigh, Sir Walter. *The History of the World*. London, 1617.

Rastell, John. *Three Rastell Plays: "Four Elements," "Calisto and Melebea," "Gentleness and Nobility,"* ed. Richard Axton. Cambridge , England: D. S. Brewer, 1979.

Recorde, Robert. *The Ground of Arts*. London, 1543.
Reid, Katie. *Martianus Capella in the Late Middle Ages and Renaissance*. Medieval and Renaissance Authors and Texts, vol. 28. Leiden, the Netherlands: Brill, 2023.
Reisch, Gregor. *Natural Philosophy Epitomised: A Translation of Books 8-11 of Gregor Reisch's "Philosophical Pearl" (1503)*, ed. and trans. Andrew Cunningham and Sachiko Kusukawa. Farnham, Surrey: Ashgate, 2010.
Reiss, Timothy J. *Mirages of the Selfe: Patterns of Personhood in Ancient and Early Modern Europe*. Stanford, CA: Stanford University Press, 2003.
Remien, Peter. *The Concept of Nature in Early Modern English Literature*. Cambridge: Cambridge University Press, 2019.
Remmelin, Johann. *Catoptrum Microcosmicum*. Ulm, Germany, 1639.
Remmert, Volker R. *Picturing the Scientific Revolution*, trans. Ben Kern. Philadelphia, PA: Saint Joseph's University Press, 2011.
Roberts, Michael. "Genesis and Geology UnEarthed." *The Churchman*, vol. 112, no. 3 (1998): 225–55.
Roberts, Michael. "The Genesis of John Ray and His Successors." *Evangelical Quarterly*, vol. 74, no. 2 (2002): 143–63.
Robinson, Ken. "The Skepticism of Butler's Satire on Science: Optimistic or Pessimistic?" *Restoration: Studies in English Literary Culture, 1660–1700*, vol. 7, no. 1 (1983): 1–7.
Rogers, John. *The Matter of Revolution: Science, Poetry, and Politics in the Age of Milton*. Ithaca, NY: Cornell University Press, 1996.
Rosenberg, Daniel, and Anthony Grafton. *Cartographies of Time: A History of the Timeline*. Princeton, NJ: Princeton Architectural Press, 2013.
Rosenthal, Earl. "Plus Ultra, Non Plus Ultra, and the Columnar Device of Emperor Charles V." *Journal of the Warburg and Courtauld Institutes*, vol. 34, no. 1 (1971): 204–28.
Rossi, Paolo. *Francis Bacon: From Magic to Science*, trans. Sacha Rabinovitch. Chicago, IL: University of Chicago Press, 1968.
Rowe, Katherine. "'God's Handy Worke': Divine Complicity and the Anatomist's Touch." In *The Body in Parts: Fantasies of Corporeality in Early Modern Europe*, ed. David Hillman and Carla Mazzio. New York: Routledge, 1997: 285–309.
Rudwick, Martin J. S. *Earth's Deep History: How It Was Discovered and Why It Matters*. Chicago, IL: University of Chicago Press, 2014.
Ruse, Michael. "Creation Science Is Not Science." *Science, Technology, & Human Values*, vol. 7, no. 40 (1982): 72–8.
Sagan, Carl. *Pale Blue Dot: A Vision of the Human Future in Space*. New York: Random House, 1994.
San Juan, Rose Marie. "The Turn of the Skull: Andreas Vesalius and the Early Modern Memento Mori." *Art History*, vol. 35, no. 5 (2012): 958–75.

Sarasohn, Lisa T. "Fideism, Negative Theology, and Christianity in the Thought of Margaret Cavendish." In *God and Nature in the Thought of Margaret Cavendish*, ed. Brandie R. Siegfried and Lisa T. Sarasohn. London: Routledge, 2014: 93–106.
Sarasohn, Lisa T. *The Natural Philosophy of Margaret Cavendish: Reason and Fancy during the Scientific Revolution*. Baltimore, MD: Johns Hopkins University Press, 2010.
Sarkar, Debapriya. *Possible Knowledge: The Literary Forms of Early Modern Science*. Philadelphia: University of Pennsylvania Press, 2003.
Sawday, Jonathan. *The Body Emblazoned: Dissection and the Human Body in Renaissance Culture*. London: Routledge, 1995.
Schabas, Margaret. *The Natural Origins of Economics*. Chicago, IL: University of Chicago Press, 2005.
Schabas, Margaret, and Neil de Marchi. *Oeconomies in the Age of Newton*. Durham, NC: Duke University Press, 2003.
Schaff, Philip, ed. *A Select Library of the Nicene and Post-Nicene Fathers of the Christian Church: Vol. II: St. Augustine's "City of God" and Christian Doctrine*. Buffalo, NY: The Christian Literature Co., 1887.
Schlick, Moritz. *Philosophy of Nature*, trans. Amethe von Zeppelin. New York: Greenwood Press, 1968.
Scott, Walter, ed. *Hermetica: The Ancient Greek and Latin Writings*, vol. 1. Boston, MA: Shambhala, 1985.
Seelig, Sharon Cadman. *The Shadow of Eternity: Belief and Structure in Herbert, Vaughan, and Traherne*. Lexington: University of Kentucky Press, 1981.
Seidman, Naomi. *Faithful Renderings: Jewish–Christian Difference and the Politics of Translation*. Chicago, IL: University of Chicago Press, 2006.
Shakespeare, William. *The Complete Works*, ed. Stanley Wells and Gary Taylor. Oxford: Clarendon Press, 1988.
Shakespeare, William. *Hamlet: A Norton Critical Edition*, ed. Robert Miola. New York: Norton, 2019.
Shakespeare, William. *The Norton Shakespeare*, ed. Stephen Greenblatt, Walter Cohen, Suzanne Gossett, Jean E. Howard, Katharine E. Maus, and Gordon McMullan. 3rd ed. New York: W. W. Norton, 2016.
Shapin, Steven. *The Scientific Revolution*. Chicago, IL: University of Chicago Press, 1996.
Shapin, Steven. *A Social History of Truth: Civility and Science in Seventeenth-Century England*. Chicago, IL: Chicago University Press, 1994.
Shapin, Steven. "Understanding the Merton Thesis." *Isis*, vol. 79, no. 4 (1988): 594–605.
Shapin, Steven, and Simon Schaffer. *Leviathan and the Air-Pump: Hobbes, Boyle, and the Experimental Life*. Princeton, NJ: Princeton University Press, 2011.
Sherman, Donovan. *Second Death: Theatricalities of the Soul in Shakespeare's Drama*. Edinburgh: Edinburgh University Press, 2016.
Sidney, Sir Philip. *An Apology for Poetrie*, ed. Evelyn S. Shuckburgh. Cambridge: Cambridge University Press, 1951.

Sidney, Sir Philip. "The Defence of Poesy." In *The Major Works*, ed Katherine Duncan-Jones. Oxford: Oxford University Press, 2008: 212–51.

Siegfried, Brandie R., and Lisa T. Sarasohn, eds. *God and Nature in the Thought of Margaret Cavendish*. Farnham, Surrey: Ashgate, 2014.

Sigelman, Aysa C. *Pindar's Poetics of Immortality*. Cambridge: Cambridge University Press, 2016.

Siraisi, Nancy G. *History, Medicine, and the Traditions of Renaissance Learning*. Ann Arbor: University of Michigan Press, 2008.

Sissa, Guilia. *Greek Virginity*, trans. Arthur Goldhammer. Cambridge, MA: Harvard University Press, 1990.

Sissa, Guilia. "The Hymen Is a Problem, Still: Virginity, Imperforation, and Contraception, from Greece to Rome." *EuGeStA: Journal of Gender Studies in Antiquity*, vol. 3 (2013): 67–123.

Sissa, Guilia. "Maidenhood Without Maidenhead: The Female Body in Ancient Greece." In *Before Sexuality: The Construction of Erotic Experience in the Ancient Greek World*, ed. D. M. Halperin, J. C. Winkler, and F. J. Zeitlin. Princeton, NJ: Princeton University Press, 1990: 339–64.

Sissa, Guilia. "Subtle Bodies." In *Fragments for a History of the Human Body, Part Three*, ed. M. Feher, R. Naddoff, and Nadia Tazi. New York: Urzone, 1989: 132–57.

Skouen, Tina. "Brian Cox and the New Enlightenment." *Notes and Records of the Royal Society*, vol. 74, no. 1 (2020): 149–78.

Skouen, Tina. "Science Versus Rhetoric? Sprat's *History of the Royal Society* Reconsidered." *Rhetorica: A Journal of the History of Rhetoric*, vol. 29, no. 1 (2011): 23–52.

Sloane, Hans. *A Voyage to the Islands Madera, Barbados, Nieves, S. Christophers and Jamaica*. 2 vols. London, 1707.

Smith, Courtney Weiss. *Empiricist Devotions: Science, Religion, and Poetry in Early Eighteenth-Century England*. Charlottesville: University of Virginia Press, 2016.

Smith, Hilda L. "Claims to Orthodoxy: How Far Can We Trust Margaret Cavendish's Autobiography?" In *God and Nature in the Thought of Margaret Cavendish*, ed. Brandie R. Siegfried and Lisa T. Sarasohn. London: Routledge, 2014: 15–26.

Soranus of Ephesus. *Gynecology*, trans. Oswei Temkin. Baltimore, MD: Johns Hopkins Press, 1956.

Soranus of Ephesus. *Soranos D'Ephèse, Maladies des Femmes*, ed. and trans. P. Burgière, D. Gourevitch, Y. Malinas. 4 vols. Paris: Les Belles Lettres, 1988.

Spiller, Elizabeth. *Science, Reading, and Renaissance Literature: The Art of Making Knowledge, 1580–1670*. Cambridge: Cambridge University Press, 2004.

Sprat, Thomas. *The History of the Royal Society of London for the Improving of Natural Knowledge*. London, 1667.

Sprat, Thomas. *The History of the Royal Society*, ed. Jackson I. Cope and Harold Whitmore Jones. London: Routledge, 1966.

Spruit, Leen. *Species Intelligibilis: From Perception to Knowledge: Vol. 2: Renaissance Controversies, Later Scholasticism, and the Elimination of the Intelligible Species in Modern Philosophy*. Leiden, the Netherlands: Brill, 1995.

Steiner, Gary. "The Cultural Significance of Rembrandt's Anatomy Lesson of Dr. Nicolaas Tulp." *History of European Ideas*, vol. 36, no. 3 (2010): 273–9.

Stillman, Robert. "The State (Out) of Language: Dryden's 'Annus Mirabilis' as a Restoration Paradigm for Scientific Revolution." *Soundings: An Interdisciplinary Journal*, vol. 84, no. 1/2 (2001): 201–27.

Strickland, Lloyd. "The Doctrine of 'The Resurrection of the Same Body' in Early Modern Thought." *Religious Studies*, vol. 46, no. 2 (2010): 163–83.

Stump, Eleonore. *Aquinas*. London: Routledge, 2003.

Sugg, Richard. *Murder After Death: Literature and Anatomy in Early Modern England*. Ithaca, NY: Cornell University Press, 2007.

Sugg, Richard. *The Secret History of the Soul: Physiology, Magic, and Spirit Forces from Homer to St. Paul*. Cambridge: Cambridge Scholars, 2013.

Sugg, Richard. *The Smoke of the Soul: Medicine, Physiology and Religion in Early Modern England*. New York: Palgrave Macmillan, 2013.

Summit, Jennifer. *Lost Property: The Woman Writer and English Literary History, 1380–1589*. Chicago, IL: The University of Chicago Press, 2000.

Swann, Elizabeth L., and Subha Mukherji, eds. *The Poesy of "Scientia" in Early Modern England*. London: Palgrave, 2024.

Swift, Jonathan. *Gulliver's Travels*, ed. Albert J. Rivero. New York and London: W. W. Norton, 2002.

Toulmin, Stephen, and June Goodfield. *The Discovery of Time*. Chicago, IL: University of Chicago Press, 1965.

Traherne, Thomas. *Selected Poems and Prose*, ed. Alan Bradford. New York: Penguin Books, 1991.

Traub, Valerie. "Gendering Mortality in Early Modern Anatomies." In *Feminist Readings of Early Modern Culture: Emerging Subjects*, ed. Valerie Traub, M. Lindsay Kaplan, and Dympna Callaghan. Cambridge: Cambridge University Press, 1996: 44–92.

Traub, Valerie. "The Nature of Norms in Early Modern England: Anatomy, Cartography, *King Lear*." In "Shakespeare and Science, c. 1600," *South Central Review*, vol. 26, no. 1 & 2 (Spring & Summer 2009): 41–82.

Trollinger, Susan L. and William Vance Trollinger, Jr. *Righting America at the Creation Museum*. Baltimore, MD: Johns Hopkins University Press, 2016.

Ungureanu, James C. "Relocating the Conflict Between Science and Religion at the Foundations of the History of Science." *Zygon*, vol. 53, no. 4 (2018): 1106–30.

Ungureanu, James C. *Science, Religion, and the Protestant Tradition: Retracing the Origins of Conflict*. Pittsburgh, PA: University of Pittsburgh Press, 2019.

Ussher, James. *Annales veteris testamenti*. London, 1650.

Ussher, James. *The Annals of the World Deduced from the Origin of Time*. London, 1658.

Vaughan, Henry. *The Works of Henry Vaughan*, ed. L. C. Martin. Oxford: Clarendon Press, 1957.

Veith, Ilza. *Hysteria: The History of a Disease*. Chicago, IL: University of Chicago Press, 1965.

Vesalius, Andreas. *The China Root Epistle: A New Translation and Critical Edition*, ed. and trans. D. H. Garrison. Cambridge: Cambridge University Press, 2015.

Vesalius, Andreas. *De humani corporis fabrica libri septem*. Basel, 1543.

Vesalius, Andreas. *De humani corporis fabrica librorum epitome*. Basel, 1543.

Vesalius, Andreas. *The Epitome of Andreas Vesalius*, trans. L. R. Lind, ed. C. W. Asling and Logan Clendening. New York: Macmillan, 1949.

Vesalius, Andreas. *The Fabric of the Human Body: An Annotated Translation of the 1543 and 1555 Editions*, ed. and trans. D. H. Garrison and M. H. Hast. 2 vols. Basel: Karger, 2014.

Vesalius, Andreas. *On the Fabric of the Human Body: A Translation of "De Humani Corporis Fabrica Libri Septem"*, ed. and trans. William Frank Richardson and John Burd Carman. 5 vols. Novato, CA: Norman Publishing, 1998–2009.

Vickers, Brian. "Francis Bacon, Feminist Historiography, and the Dominion of Nature." *Journal of the History of Ideas*, vol. 69, no. 1 (2008): 117–41.

Vickers, Brian. "The Royal Society and English Prose Style: A Reassessment." In *Rhetoric and the Pursuit of Truth: Language Change in the Seventeenth and Eighteenth Centuries*, ed. Brian Vickers and Nancy S. Struever. Los Angeles: William Andrews Clark Memorial Library for the University of California, Los Angeles Press, 1985: 1–76.

Vidal, Fernando. *The Sciences of the Soul: The Early Modern Origins of Psychology*, trans. Saskia Brown. Chicago, IL: University of Chicago Press, 2006.

Von Gunten, Hans R. "Radioactivity: A Tool to Explore the Past." *Radiochimica Acta*, vol. 70, no. 71 (1995): 305–16.

Von Staden, Heinrich. "The Discovery of the Body: Human Dissection and Its Cultural Contexts in Ancient Greece." *Yale Journal of Biology and Medicine*, vol. 65, no. 3 (1992): 223–41.

Walker, D. P. *The Ancient Theology: Studies in Christian Platonism from the Fifteenth to the Eighteenth Century*. Ithaca, NY: Cornell University Press, 1972.

Walker, D. P. "The *Prisca Theologia* in France." *Journal of the Warburg and Courtauld Institutes*, vol. 17, no. 3/4 (1954): 204–59.

Walker, Julia M., ed. *Dissing Elizabeth: Negative Representations of Gloriana*. Durham, NC: Duke University Press, 1998.

Wallis, John. *De sectionibus conicis nova methodo expositis tractatus*. Oxford, 1655.

Wang, Weija. "Kant's Mathematical Sublime: The Absolutely Great in Aesthetic Estimation." *Kantian Review,* vol. 25, no. 3 (2020): 465–85.
Weber, Max. *The Protestant Ethic and the Spirit of Capitalism*, trans. Stephen Kalberg. New York: Routledge, 2012.
Weber, Max. "Science as a Vocation." In *The Vocation Lectures*, ed. David S. Owen and Tracy B. Strong, trans. Rodney Livingston. Indianapolis, IN: Hackett, 2004: 1–31.
Weber, Max. *The Vocation Lectures*, trans. Rodney Livingston, ed. David S. Owen and Tracy B. Strong. Indianapolis, IN: Hackett, 2004.
Werth, Tiffany Jo. "Atheist, Adulterer, Sodomite, Thief, Murderer, Lyer, Perjurer, Witch, Conjuror, Brute Beast? Discovering the Ungodly in Shakespeare's England." *Literature Compass*, vol. 10, no. 2 (2013): 175–88.
Westfall, Richard. *Never at Rest: A Biography of Isaac Newton*. United Kingdom: Cambridge University Press, 1983.
White, Andrew Dickson. *A History of the Warfare of Science with Theology in Christendom*, 2 vols. New York: D. Appleton, 1896.
Wilkins, John. *An Essay Towards a Real Character and a Philosophical Language*. London, 1668.
Willis, Thomas. *Two Discourses Concerning the Souls of Brutes*, trans. S. Pordage. London, 1683.
Willsford, Thomas. *Willsford's Arithmetic, Natural and Artificial*. London, 1656.
Witmore, Michael. *Culture of Accidents: Unexpected Knowledges in Early Modern England*. Stanford, CA: Stanford University Press, 2001.
Wittgenstein, Ludwig. *Philosophical Investigations*, trans. G. E. M. Anscombe, Joachim Schulte, and P. M. S. Hacker. 4th ed. Malden, MA: Wiley Blackwell, 2009.
Wofford, Susanne L. "Foreign." In *Early Modern Theatricality*, ed. Henry S. Turner. Oxford: Oxford University Press, 2013: 478–92.
Wofford, Susanne L. "Hymen and the Gods Onstage in Shakespeare's *As You Like It* and Italian Pastoral." In *Transnational Mobilities in Early Modern Theater*, ed. Robert Henke and Eric Nicholson. Aldershot, Hampshire: Ashgate, 2014: 69–92.
Wood, Anthony à. *Athenæ Oxoniensis*. 2 vols. London, 1721.
Woolf, Virginia. *A Writer's Diary*. London: Harcourt, 2003.
Worster, Donald. *Nature's Economy: A History of Ecological Ideas*. Cambridge: Cambridge University Press, 1992.
Wright, Gillian. "Nature in Cowley." In *Words at War: The Contested Language of the English Civil War*, ed. Paul Hammond and Andrew Hadfield. London: British Academy, 2024: 145–57.
Wroth, W. W., and Philip Carter. "Hales, William (1747–1831), Church of Ireland Clergyman and Scientific Writer." *Oxford Dictionary of National Biography*. September 23, 2004.
Yates, Frances. *Giordano Bruno and the Hermetic Tradition*. Chicago, IL: University of Chicago Press, 1964.

Yates, Frances. "The Hermetic Tradition in Renaissance Science." In *Art, Science, and History in the Renaissance*, ed. Charles Singleton. Baltimore, MD: Johns Hopkins University Press, 1967: 255–76.

Yelle, Robert A., and Lutz Trein, eds. *Narratives of Disenchantment and Secularization: Critiquing Max Weber's Idea of Modernity*. London: Bloomsbury, 2021.

Zagorin, Peter. *Francis Bacon*. Princeton, NJ: Princeton University Press, 1998.

Zimmerman, Michael. *Heidegger's Confrontation with Modernity: Technology, Politics, and Art*. Bloomington: Indiana University Press, 1990.

Index

www.ingramcontent.com/pod-product-compliance
Ingram Content Group UK Ltd.
Pitfield, Milton Keynes, MK11 3LW, UK
UKHW020904190926
475560UK00010B/188

9 781350 511620